2017 coH 404 July 2nd

Breezy and free! 1½ dogs, pine next door on marsh edge, FULL.

Joey! roseate spoonbill over marsh

(never saw Bruce's eagle nest on causeway)

Pooh's Owl great horned, at Elbow Bend .. + all the usual suspects

Research spotting nests (2) of, previous rental- Joey + Bill pointed out the → Osprey's at Elbow Bend long ago Joey spotted + showed me a Indigo Bunting "Like a quilt"?

Jim Wilson

Common
Birds

OF COASTAL GEORGIA

The University of Georgia Press
Athens and London

Published by the University of Georgia Press

Athens, Georgia 30602

www.ugapress.org

Designed by Mindy Basinger Hill

Set in 10.5/14 pt Garamond Premier Pro

Printed and bound by Imago

The paper in this book meets the guidelines for
permanence and durability of the Committee on
Production Guidelines for Book Longevity of the
Council on Library Resources.

Printed in China

15 14 13 12 11 P 5 4 3 2 1

Library of Congress Cataloging-in-Publication Data

Wilson, Jim, 1944 Dec. 21–

Common birds of coastal Georgia / Jim Wilson.

        p.   cm. — (A Wormsloe Foundation nature book)

Includes index.

ISBN-13: 978-0-8203-3828-6 (pbk. : alk. paper)

ISBN-10: 0-8203-3828-1 (pbk. : alk. paper)

1. Birds —Georgia —Atlantic Coast Region.

2. Birds —Georgia —Atlantic Coast Region —Identification.

I. Title.

QL684.G4W547 2011

598.09758 —dc22   2010032580

British Library Cataloging-in-Publication Data available

DEDICATED TO ANSELM ATKINS

NATURALIST, WRITER, ARTIST,

AND FRIEND

Contents

Ocean and Shore

Acknowledgments

I am grateful for the help I received from many sources. Thanks go to all the birders who allowed me to set up and photograph birds in their backyards. In particular, Art and Barbara English, Earl Horn, Melanie Haire, Rusty Trump, Diane Barnsley, John Johnson, Ida Kinney, Charlie Muise, Beth Roth, Paul Reed, Linda Wilson, Thad Weed, Dot Bambach, Rene Heidt, Monteen McCord, and Skidaway Island State Park were great resources. Frank Kiernan was invaluable for his knowledge of photography, as was Chang-Kwei Lin, who kindly helped me with the computer aspects of the project. I also thank Bill Blakeslee and Giff Beaton for checking the text and photographs for accuracy, and Janette Younkin for checking grammar. However, any errors in the final version are my responsibility. Dr. Anselm Atkins wrote the text for many of the backyard birds and provided the inspiration for this book. Mike Chapman, Roger Clark, Don Cohrs, Pat Metz, Peggy Powell, Lydia Thompson, and Brad Winn graciously reviewed, discussed, and helped me select the species that are most common on the coast. Appreciations go to the staff at the University of Georgia Press who kindly helped steer me through the publishing process, particularly Nicole Mitchell, who decided this book was worth publishing, Judy Purdy, Mindy Basinger Hill, Jon Davies, and Walton Harris, as well as to my copyeditor Mindy Conner.

Once again, my wife, Kay, was patiently tolerant of the time I spent on this book rather than other work, and I cannot thank her enough.

Introduction

"What bird is that in my yard, or at the beach, or in the marsh?" If you live along the Georgia coast or just about anywhere else along the southern part of the Atlantic Coast, you can probably answer that question by looking here. This book provides descriptions of more than 100 of the most common birds—the ones you're most likely to see—and gives you the means to make a quick identification. Large photographs allow you to match the description with the bird you are watching. Brief, informative text tells you what to observe as you watch your bird and provides other notable facts about the species.

Bird-watching is easy, inexpensive, and fun for adults and children, so don't be surprised if you become hooked. It is indeed a hobby for a lifetime, admitting innumerable levels of interest, commitment, and reward. One reason watching birds is such an enjoyable experience is that anyone can do it, with or without binoculars, anywhere—whether at home, at work, or on vacation. You can do it while working in your garden, eating a meal, puttering in your workshop, or sitting and relaxing. Just look out the window and there they are! If you put up a feeder, you can bring the birds even closer to you—close enough, in fact, to obtain good photographs with a small camera. Also, bird behavior is fascinating to observe, whether you're viewing close interactions within a particular species or fights between various species over food and territory. Once you identify a bird, you'll want to know whether it is male or female, young or old, and whether it is a resident, a migrant, or an unusual species—and that's why I wrote this book.

HOW TO USE THIS BOOK

The birds in this book are arranged into three groups: backyard birds, marsh and pond birds, and ocean and shore birds. The birds within each group are arranged by size, from the tiny ruby-throated hummingbird to the turkey vulture in your backyard, from the common yellowthroat to the great blue heron in the marsh, and from the least sandpiper to the brown pelican at the beach. If you go first to habitat and then to size, you won't waste time searching the wrong part of the book.

Many bird species live here all year, but many others are seen or heard only during their breeding times in the spring and/or summer, in the winter when they move down from the cold north, or in the spring or fall as they pass through on their way north or south. The birds selected to appear here are those most likely to be seen or heard by residents of the coast going about their daily activities. Even the experts do not agree, so take the selections with a grain of salt.

To use the book, start with the color photographs. These have been selected to present the aspects of the bird that are most important in identification. With the Carolina wren, for example, it was important to show the white eye stripe, the buffy breast, and the typically upturned tail. If the male and female of a species are markedly different (as is the case for the northern cardinal), both sexes are shown. Sometimes two views of a bird were necessary to show all the important features or simply to provide another angle. The size of a bird is the length from the tip of its beak to the end of its tail.

The text tells you what the bird is supposed to look like, its preferred habitat, its usual food, its nesting habits, and other interesting facts. Usually, something is said of its song as well, because although birds don't sing all the time—and some rarely sing—their song is an important identifying feature if you happen to hear it.

THE NEXT STEP IN ENJOYING BIRDS

There are many ways for bird-watchers to enrich the experience. Owning this book is a first step beyond the thrill of simply watching by allowing you to actually identify the birds you see. What else can you do to increase your enjoyment?

Equipment. Binoculars are indispensable for most bird identification. Start with an inexpensive pair of 7 × 35 binoculars. By the time you want something better, you'll know what to buy. As you grow more experienced you may want a more complete field guide (see the list on pages xviii–xix). To go with your field guides, you may want to obtain some tapes or CDs of birdsongs. Much of your experience with birds will probably be listening alone.

Web Sites. There are several good Web sites and discussion groups that relate to birds. Georgia Birders Online is a good first stop, as it lets everyone across the state know what is being seen and where to go for many species. Also, there are hotlines that can be called for rare bird alerts (see Coastal Georgia Bird Information, below). Many Audubon chapters have their own Web sites with information about local activities. Go first to the National Audubon site (www.nas.org) and then to the local ones from there. The Georgia Ornithological Society (GOS) is also a good source for birding information (www.gos.org).

Field Trips. Audubon chapters sponsor group bird walks to see both resident and migratory birds, common and rare. *Common Birds of Coastal Georgia* does not cover all migrants, but by taking advantage of these free field trips, you can observe hundreds of species of birds.

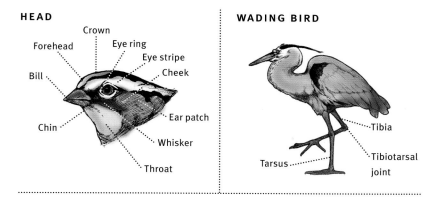

HEAD

Crown
Forehead
Eye ring
Eye stripe
Bill
Cheek
Chin
Ear patch
Whisker
Throat

WADING BIRD

Tibia
Tarsus
Tibiotarsal joint

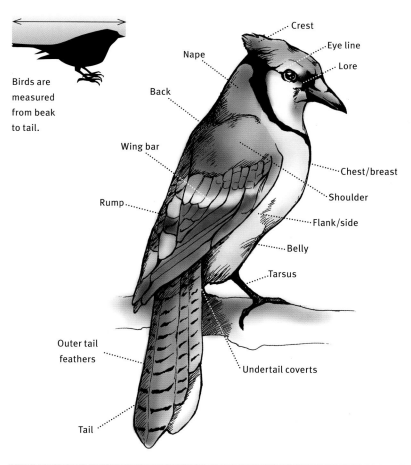

Birds are measured from beak to tail.

Crest
Eye line
Nape
Lore
Back
Wing bar
Chest/breast
Shoulder
Rump
Flank/side
Belly
Tarsus
Outer tail feathers
Undertail coverts
Tail

Bird Parts Mentioned in Species Accounts

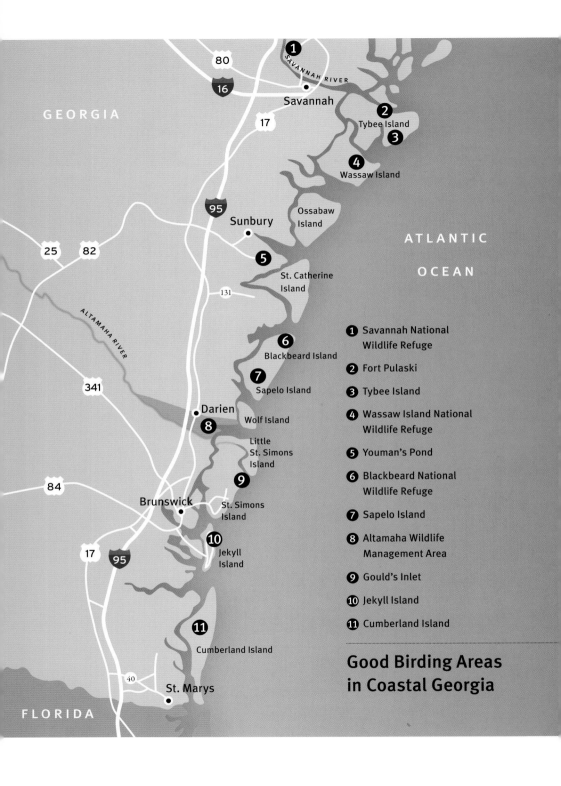

Good Birding Areas in Coastal Georgia

1. Savannah National Wildlife Refuge
2. Fort Pulaski
3. Tybee Island
4. Wassaw Island National Wildlife Refuge
5. Youman's Pond
6. Blackbeard National Wildlife Refuge
7. Sapelo Island
8. Altamaha Wildlife Management Area
9. Gould's Inlet
10. Jekyll Island
11. Cumberland Island

Christmas Bird Count. Every December, the Audubon chapters in the state sponsor holiday bird counts in which bird lovers spend the day identifying and recording every single bird seen or heard within a specific area. The atmosphere is fun and casual, but the results are important. CBC volunteers have made it possible to compile nationwide statistics on the health or decline of the majority of our species. To participate in a CBC, contact your nearest Audubon chapter.

Feeder Watch. Cornell University sponsors a nationwide bird count that lets backyard bird-watchers like you make a small contribution to bird research. From November through March, observers set aside two days every other week and spend whatever time is convenient observing the birds at their feeders. Cornell also conducts a watch day for migratory birds in May. To sign up, contact Cornell Lab of Ornithology, 159 Sapsucker Woods Road, Ithaca, NY 14850, or http://birds.cornell.edu.

Birdathon. Many Audubon Society chapters sponsor an annual birdathon as a money-raising event. Birders find sponsors (often at the workplace) who will donate so much per bird species seen. You have a wonderful day birding—and Audubon makes money!

Checklists. A checklist is a list of all the birds that have been observed in a designated place—perhaps in the state, city, wildlife refuge, or nature center. The list tells you not only what species might be seen but also what times of year each species can be found and in what numbers (common, rare, etc.). It may also tell you whether they nest there.

Listing. You might want to keep a life list: a list of all the bird species you have identified over the years. The United States alone has more than 700 species of birds, and there are about 10,000 in the world. You can keep a yard list too: species seen or heard in your own yard, which could be well over 100 species! Some enthusiastic birders keep state or even county lists.

BIRDS IN PERIL

All is not well in the avian world. Humans are hard, vigorous users of the world and its resources, and our activities are causing the decline or disappearance of many other species. We wish no harm to the coinhabitants of our planet, but when we need or want what they have, we take it. Usually what we take is their habitat. Indeed, our appetite for the habitats of plants and other animals has been ravenous. The decline in duck populations, for example, is due not to duck hunters but to law-abiding farmers who plow their wetlands under so humans can have more food. More human mouths to feed mean less habitat for ducks. Unless we can devise ways of preserving animals' space, they will, species by species, go the way of the dodo. An important step in such preservation is to stop our human population overgrowth.

The abundance of birds is an indication of how humans are affecting bird, plant, and animal habitats. As the equatorial rain forests of the world are leveled at a galloping rate, the migratory birds that winter there dwindle. As wooded nesting areas in North America are chopped up for human development, these same birds dwindle all the more. And people who love birds worry about it.

People who love humanity worry about it too, because birds are an indicator of the general health of our environment as a whole. Like the miner's canary, our everyday birds, by their disappearance, often give the first warning sign of something gone awry with our own human habitat. When bald eagles, ospreys, and peregrine falcons were disappearing because widespread use of DDT was causing their eggs to break and fail, it was a clear sign that too many harmful chemicals were loose in our land. The indiscriminate release of various chemicals into our surroundings continues, and it can only harm all of us in the end.

It is for this reason that your local Audubon Society and many other environmental organizations not only encourage and help you with your interest in birds, but also remind you that good conservation habits are essential if we are to continue to live in a world friendly to both birds and humans. As an example, if birds are hitting your windows, use ultraviolet stickers to alert them to this danger; birds have an extra photopigment that allows them to see such stickers as if they were neon signs, making for a quick and easy solution that helps conserve our birds.

OTHER GOOD BIRD BOOKS

As your interest in birds increases, you'll want to consult more detailed guides to bird identification as well as general introductions to the study of birds. A few of the best are listed below.

Erlich, Paul, David Dobkin, and Darryl Wheye. *The Birder's Handbook: A Field Guide to the Natural History of North American Birds*. New York: Simon and Schuster, 1988. A comprehensive handbook suitable for advanced birders.

Kaufman, Kenn. *Birds of North America*. Boston: Houghton Mifflin, 2000. Good all-around identification field guide using photographs.

National Geographic Society Field Guide to the Birds of North America, 2nd ed. Washington, D.C.: National Geographic Society, 1987. Paintings of birds with descriptions and range maps on the same page. Covers all birds north of Mexico, including accidental, vagrant, and casual species.

Pasquier, Roger F. *Watching Birds: An Introduction to Ornithology*. Boston: Houghton Mifflin, 1980. A good general introduction to the subject.

Peterson, Roger Tory. *A Field Guide to the Birds of Eastern and Central North America*. 5th ed. Boston: Houghton Mifflin, 2002. Recommended for beginning and intermediate birders. Field marks make identification quick and certain.

Robbins, Chandler S., Bertel Bruun, and Herbert S. Zim. *Birds of North America*. New York: Golden Press, 1966. Pictures, maps, and sonagrams are included; does not use field marks.

Sibley, David A. *The Sibley Guide to Birds of Eastern North America*. New York: Alfred A. Knopf/Chanticleer Press, 2003. Excellent coverage of various plumages of 650 North American bird species east of the Rocky Mountains; uses drawings to illustrate the differences. Very useful for identification of male, female, and immature birds in all seasons.

Stokes, Donald, and Lillian Stokes. *Field Guide to Birds: Eastern Region*. Boston: Little, Brown, 1996. Excellent photographs show different plumages and most field marks. All information is on one page, including up-to-date range map, behavior, and conservation status.

COASTAL GEORGIA BIRD INFORMATION

Many organizations contribute to the study and conservation of the birds found along the coast of Georgia. Names and other information on some of these are listed below for the benefit of concerned citizens who might wish to enlist their help, join, or otherwise contact them.

Animal rehabilitation centers: www.southeasternoutdoors.com/rehab_usa.html

Coastal Georgia Audubon Society: www.coastalgeorgiabirding.com

Georgia Birders Online (GABO): subscribe via e-mail at GABO-L@listserv.uga.edu

Georgia Department of Natural Resources, Wildlife Resources Division: One Conservation Way, Ste. 310, Brunswick, GA 31520; 912-264-7218; http://georgiawildlife.dnr.state.ga.us

Georgia Rare Bird Alert: 770-493-8862

Georgia Ornithological Society: www.gos.org

Georgia Wildlife Federation: 11600 Hazelbrand Rd., Covington, GA 30014; 770-787-7887; www.gwf.org

Humane Society of the United States regional office: Tallahassee, Fla.; 850-386-3435

Ogeechee Audubon Society: P.O. Box 13806; Savannah, GA 31416; www.ogeecheeaudubon.org

Ruby-throated Hummingbird *Archilochus colubris*

In the eastern United States, the hummingbird we're most likely to see is the ruby-throated, but only from spring through early fall. In recent years, more and more sightings of western hummers have been reported, but these are winter birds, present after ruby-throats have migrated south. Our summer hummingbirds are almost all ruby-throats.

The male hummingbird's throat feathers, or "gorget," are red, with a black chin above. The back is green; the underparts are white. Females and juveniles lack the throat patch. Because the red on the patch is iridescent, it may appear black under certain lighting conditions.

Ruby-throats are very possessive of their food sources, which are critical for their survival. When several are present at a feeder, you'll see them chasing each other away and hear their squeaky, twittery chatter; otherwise just the hum of their rapidly beating wings can be heard. Hummingbirds can assume many acrobatic positions and can even fly backward.

Female hummingbirds make a tiny nest and feed their young by regurgitation, sticking their long bill right down into the chick's gullet.

Hummingbirds use their tubular tongue to sip nectar from long-necked flowers, preferring red and orange to other colors. They also take spiders and insects and rob sap from sapsucker holes. Because of their high metabolism and extremely rapid wing beats, hummers need to eat nearly all the time. At night, to save energy, they may fall into a deep torpor that resembles hibernation.

Home feeders are well worth the effort. Fill them with a briefly boiled sugar solution made of four parts water to one part sugar. Change the fluid at least weekly and clean the feeder at the same time. Do not add red coloring; just have something red on or near the feeder itself.

The female has white-tipped outer tail feathers.

Backyard briefly,
only in the summer

The female ruby-throat has a white throat — no gorget.

The bright red gorget of the male (above left) appears completely black when the light comes from a different angle (above right).

Ruby-crowned Kinglet *Regulus calendula*

The ruby-crowned kinglet skitters quickly through the trees searching for insects, often in the company of other small birds. One thing about him makes him special: the flaming scarlet crest, which he usually raises only when he gets in a snit with rival males or when scolding a screech-owl. Otherwise, it stays hidden below his superficial crest feathers.

This kinglet has a light breast with olive-brown or gray wings and back. Its white eye ring is conspicuous as are its white wing bars. These are excellent field marks for identification along with its small size—even smaller than a chickadee. Additional marks are a short tail and constant wing flicking. No other very small bird does this so regularly. Don't expect to see the red crest when he is not excited.

The ruby-crowned has a raised red crest when it is excited (top), but normally it is hidden (as it mostly is in the bottom image).

This kinglet is very active, darting among the bare twigs of winter trees or searching the green boughs of pines. Its food may be berries of honeysuckle and poison ivy, or whatever remains from the bounty of summer along with whatever insects are available. It continues to flick its wings in and out while foraging, so think kinglet when you see this behavior.

On the ruby-crowned's wintering grounds in the South, you may hear him scold and think at first of titmouse or wren. His song, however, a loud descending series of varied notes, is one you'll never hear on a Christmas bird count because those don't take place during the breeding season.

The ruby-crowned has a white eye ring and one white wing bar.

Carolina Chickadee *Poecile carolinensis*

A very common small, gray bird with a black cap, black bib, white cheeks, and white belly is unmistakably our chickadee. They'll come around your house if you have any trees at all, and will be at your feeder in an instant if you put out sunflower seeds. Watch how one eats. It picks through the seeds, casting away those it doesn't want, until it finds just the right one. Then the chickadee carries the seed to a nearby branch, holds it between its toes, and pecks at the husk. Soon it flies back for more. When not eating birdseed, the chickadee searches foliage for insects, often hanging upside down from a twig to see what's underneath.

The nest is a hollow excavated in rotten wood or any similar cavity, found or provided, including a nest box. There are usually six eggs, white with reddish speckles. Both parents bring food to the young.

The chickadee sings its name, as many birds do: a rather buzzy *chick-a-dee-dee-dee*. In spring it gives a four-note up-and-down whistled song: *fee-bee-fee-bay*. Chickadees also produce a high squeaking that is hard to describe.

Carolina chickadees like to hang around with tufted titmice, their close cousins; often you'll see a few of each at the same time. Chickadees will let you get fairly close. Seems like the larger the bird, the more wary it is. Perhaps the chickadee knows it's not big enough to make a mouthful.

This little bird is usually the first to come to a new feeder.

Birds are frequently described as awesome, majestic, pretty, handsome, interesting, weird, or maybe even boring or pesky. But only chickadees can be described as cute. And that is why we can't stop loving them.

Up north, the blacked-capped chickadee takes the place of the Carolina chickadee. The two species are very, very similar in appearance (the BCC has a little more white on the wings), but their vocalizations and range are different enough to earn them classification as separate species.

𝓔 bird feeder, Atlanta

Note the tiny bill and striking white cheek bordered by jet black above and below.

Blue-gray Gnatcatcher *Polioptila caerulea*

This little bird is common all along the coast year-round, but the ones you see in the winter may be different individuals from the ones you see in the summer. The gnatcatcher is easily recognized by its very small size and long tail, which, like the bird itself, is always in motion. In overall length the gnatcatcher is almost 1 inch shorter than a titmouse, and most of its 4 or so inches is tail, so the body is very small. It very much resembles a miniature mockingbird, with its sleek body, long tail typically cocked in the air, and similar color pattern—blue-gray on top and white underneath. The male has a black U across his forehead that extends onto the crown above his eyes.

The gnatcatcher nests in wooded river bottoms or open pinewoods with oak understories. The nest is a cup similar in shape to a hummingbird's but a bit larger. It typically sits on a horizontal branch 15–40 feet above the ground and, also like a hummingbird's nest, is covered by lichens that are held on by spiderwebs. Surrounded by leaves, it is well camouflaged and looks like a knob on the branch. Both parents feed the young and are constantly in motion picking insects from the surrounding tree leaves. Often, they dart out to snatch an insect that is flying by or that has been flushed from the leaves.

You can usually hear the gnatcatcher's high-pitched *zzzzeeeeee zzzzeeeee* several times a minute while it does its grocery shopping.

Although very tolerant of humans living nearby, the gnatcatcher is a bird of wooded country areas with ponds or streams, particularly in the wintertime.

The male (opposite) shows its long tail and a gray/white mockingbird pattern on its body, with no wingbars. Female (below) has a similar pattern, but without the black U above the eyes. Note the clear eye ring for both.

American Goldfinch *Spinus tristis*

The reason the American goldfinch is often called "wild canary" is evident in the male's spring breeding plumage: bright lemon yellow with contrasting black wings, tail, and cap. The female's yellow is very much paler, and she has a dull brownish gray back. In winter, both sexes are brownish gray with just the barest blush of yellow and more muted black. The wing will show one white stripe and a yellow epaulet. Goldfinches are notorious for liking thistles, an expensive seed for feeders. They seem just as happy with sunflower seeds, however, and can be found regularly at the trough alongside other finches. Their thick, stubby bill is good for crushing seeds.

The winter female is gray-brown with very little yellow. The winter male may have a little yellow on the shoulders and face and has no black cap.

Any yellow-colored warbler that looks a little like a goldfinch can be distinguished by its delicate, pointed warbler's bill (warblers eat mainly insects). A goldfinch in flight can often be identified by its undulations and call: *per-chik-o-ree*. The song is long, high, and sweet.

Goldfinches may flock by the dozens or hundreds in fields and in trees with seeds. So devoted is the goldfinch to thistles that it even makes its nest out of their down. Nesting seems to coincide with the maturation of thistles. Farmers, leave those pretty purple thistles in your pastures! During winter, the adults and fledglings may have left their thistle fields and come to your feeders, so make sure you have sunflower seeds as well as some thistle seeds out to accommodate them.

Occasional, W.

The summer male is bright lemon yellow with black cap and wings and white wing bars.

Indigo Bunting *Passerina cyanea*

The names of some birds reflect the varied and subtle colors of nature. Cerulean, lazuli, buff, bay, rose, ferruginous, rufous, ash, ruby, glaucous, chestnut, and clay are only a few examples. The male indigo bunting is truly a rare color—a deep, rich blue. The female has brown wings, a tan head, and a buffy breast with a hint of brown striping. In fall and winter the male resembles his mate except for touches of blue. The male always has a little black on his wings and tail. If you're able to view the indigo's beak, notice its finchlike thickness and its color: dark above, light below. A similar but rarer bird in the countryside is the blue grosbeak (not included here). That bird is a little larger and is a darker, duller bluish purple in good light, and it has two distinct reddish-brown wing bars.

City dwellers will have to do a little driving to reach the indigo bunting's favored habitat, for it loves the open country. Even there it is present only from spring through fall. The bunting feeds and nests in heavy undergrowth, and you'll seldom discover it on the ground. No matter; the male sings from an exposed perch, frequently near the top of a tree. You'll find buntings around brushy field edges or on small trees in the field. You can see them on wires and even on the roofs of outbuildings.

The male's song is sweet and distinctive: something like *sweet-sweet, where-where, here-here, see it—see it*. The paired notes are on different pitches, weakening and descending.

The male (top) is all blue except for black touches on his wings and tail. Note the light lower mandible. The female (bottom) has brown plumage, with wings darker than her head. Some light streaking is usually visible on the sides and upper breast. The male indigo is more likely to come to a feeder and perch near your house during its summer visit.

Painted Bunting *Passerina ciris*

The late Roger Tory Peterson called the male painted bunting "the most gaudily colored North American songbird." Indeed, the male's violet blue head and nape, vermilion red underparts, and yellow to green back and wings have a jewel-like brilliance. The red eye ring practically says, "How do you like my makeup and outfit?" The French name *nonpareil* (meaning "without equal") is well deserved. Despite their conspicuous colors, the males are not easily seen because they spend much of their time under dense foliage where their colors become shades of gray. Occasionally, however, they perch up high on an exposed limb or bush top where they sing their musical warble and display their beauty. The female, on the other hand, is dull green above and faded or light yellow to dull white below, making her inconspicuous as she visits her nest and broods her eggs. If you see a green finch, it must be a female painted bunting; there is no other finch with this coloration. Her bill is gray; the male's is black.

Painted buntings are found in the coastal plain of Georgia and upper Florida only in the summer, mostly along the coast and major rivers. In winter they migrate to southern Florida and the northwestern Caribbean. Oddly, there is a second population of painted buntings throughout much of Texas and adjacent states to the north and south, but few, if any, in Mississippi or Alabama. This group migrates mostly to Mexico and Central America.

Painted buntings forage for seeds and insects on the ground or in low shrubbery near open grassy areas. Occasionally, we get lucky and they come to a feeder for sunflower seeds.

at Foou tend circa 1990's
Joey spotted it "like a giant!"

The unmistakable male painted bunting (above) is perhaps the most colorful songbird in the United States. The female (opposite) is identified by her green back because no other bird with a finch beak has this color.

White-eyed Vireo *Vireo griseus*

When they are hidden among the tree leaves, the vireos can be difficult to distinguish from the warblers, but this vireo generally stays in the lower thickets and bushes rather than in the treetops where most warblers feed. Furthermore, the white-eyed vireo sings its song frequently, so you should learn these notes to save yourself the trouble of tracking this elusive bird to find out what it is. The song almost always has an initial *chip* at the beginning, quickly followed by several up-and-down notes and sometimes ending with another weaker *chip*—something like *chip, a-wee-oh, chip*. It's the initial *chip* that easily identifies this bird while it hides in the bushes.

Should you obtain a view, which can be done with a little persistence, the key features are the yellow "glasses" that surround the eyes and run over the lores. The irises are, of course, white. The back and wings are olive green, blending into gray at the neck. Two yellow wing bars adorn the wings. The underparts are white with yellowish flanks. The bill is thin, short, and black; the legs are blue-gray.

The white-eyed vireo feeds mostly on insects except in the fall, when it will also eat berries. This species migrates to Central America for the winter, although some remain on the Atlantic and Gulf coasts. Its preference for dense shrubbery or thickets near streams or swampy woodlands makes the white-eyed vireo a little less common than other backyard birds. Like many insect-eating birds, it is not easily attracted to a feeder.

This little bird is shy and rarely allows a good view, but the yellow "spectacles" and white iris are obvious when it does.

Chipping Sparrow *Spizella passerina*

The chipping sparrow has a light gray, unstreaked breast, but what marks it best is its broad, chestnut cap. Look also for the black stripe through the eye and the white line just above it. In winter the bird is plainer: the rusty cap is replaced with a dull brown one, the face becomes grayish brown, and the white stripe is drab.

Though chipping sparrows may come to a feeder, you're more likely to see them in flocks out in the countryside. You can see hundreds feeding in the gray winter corn stubble, hiding in a hedgerow, or feeding on the grass in the front yards of homes. Their flocks are typically larger than those of the field sparrow.

The chippy's conical bill marks it as a seed crusher, and seeds are what it's hunting in the fields. It eats insects too, of course. Most perching birds feed their young insects, and the chippy is no different in that regard.

The chippy's song is a rattle or trill, all on one note. It seems quieter, drier, and faster than the trill of the pine warbler, with whose song it might be confused. You may need practice to keep from confusing the two songs. Both species are present and singing during our Christmas bird counts, and likewise in the spring and early summer.

The reddish cap (top) is at its best during the summer breeding season. Also note the black stripe through the eye and the white stripe above it. The rusty cap is replaced with brown during the winter (bottom), and the eye stripe is no longer white.

The lack of a complete eye ring helps to distinguish
the chipping from the field sparrow.

Yellow-rumped Warbler *Dendroica coronata*

The yellow-rumped is best noticed and identified by the yellowish splotch of feathers above its tail. Old-time birders lovingly call it the "butter-butt." It is present in Georgia from November through May, with many individuals spending the winter here and very large numbers migrating through the state in spring and fall. The highest winter concentrations occur along the coast.

In winter plumage, the bird is drab and nondescript, though the yellow rump and a tinge of yellow on the upper breast near the wing remain. But the male in spring breeding plumage is an eyeful with his black eye patch and black pigment on the breast, yellow on the crown, and brighter yellow on the flanks and rump. The western form of this species sports a yellow throat instead of a white one.

The yellow-rump feeds largely on insects, even taking them on the wing like a flycatcher. In winter, it resorts to berries. The song is a fluctuating trill; the call, a *check*. It is often found in flocks in trees or high brush. At those times you can hear its constant little *check* call.

The yellow-rumped warbler is especially interesting because it has two distinct forms, eastern and western, that are an example of evolution in the actual process of dividing one species into two. The two subspecies can still interbreed when they meet, and thus are still considered a single species. Given enough time, however, they could very well become two different warblers. Of course, this has already happened with our other 40 or so species of warblers, which arose from a single common ancestor. Our eastern form was formerly called the "myrtle warbler," and the western form "Audubon's warbler," as if they were indeed two species, until ornithologists noticed that they were still interbreeding where their ranges overlapped. A very similar cousin, the magnolia warbler, has already gone its own way.

The female and the winter male are indistinguishable (above):
dull and brownish, but still with a yellow rump and a little
yellow wash below the bend of the wing.

Male (left) acquiring spring breeding plumage: lots of black,
some white, and bits of yellow.

Carolina Wren *Thryothorus ludovicianus*

This little brown wren is among our most lovable birds. It is distinguished from other birds by its uptilted tail, and from other wrens seen on the coast (house wren, winter wren, and marsh wren) by its prominent white eye stripe and buffy breast. Although it is larger than other eastern wrens, it is still less than 5 inches long. Its range includes the whole Southeast.

Carolina wrens will nest almost anywhere. A pair may bring leaves into an old hat hanging in your garage or to your hanging planter. They've even been known to build a nest under the hood of a car! They accept any protected place and seem to like being near human habitation. Each pair raises several broods each year. You can often see a little band of four or five wrens noisily flitting through the brush around your yard.

The Carolina wren lives mainly on insects, often hunting low to the ground. It hops briskly over brush piles, tree stumps, and shrubs, searching every cranny. But it will also come to a feeder, where it prefers to eat alone. It seems to like food particles left behind by the other birds.

The Carolina wren has an amazing repertoire boasting at least 125 distinguishable songs. The bright, cheery, rolling songs are much louder than the singer's size would suggest. The notes are in groups of threes or fours. You'll often hear *teacher, teacher, teacher* or *tea-kettle, tea-kettle, tea-kettle*. The wren will also sing triplets of *twinkie, courtesy, piecemeal, trilogy,* and many more "words." These aren't random sounds, though; they're particular songs. This wren also has a fussy hissing scold, a dry rattling "answering" trill, and a repeated bright rolling trill. Sometimes this wonderful little bird seems responsible for half the birdsong in your yard.

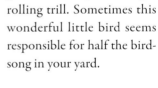

The long, wickedly curved bill; strong white eye stripe; and buffy breast are diagnostic for this species (left). The cocked tail (opposite page) is also typical. Males and females are alike.

Pine Warbler *Dendroica pinus*

It's a dull winter day, you're looking at gray clouds behind a mask of green pines, and suddenly you see a bright spot of yellow. There's only one bird this happy little spot could be: the pine warbler. Just to be sure, you check for field marks: two white wing bars; largish for a warbler, with a stronger-looking beak than usual. Then you confirm: greenish olive above, white toward the back of a yellow belly, a faint yellow eye ring. And just then he lets loose a trilling warble that settles the matter.

Be aware, though, that not all pine warblers are as yellow as the one you see. Females and immatures are duller, and even males can vary. But on your Christmas bird count, even the faintest hint of yellow will be enough to give the pine warbler away.

Pine warblers are in our area year-round. They can be fairly nondescript in the fall, but you can certainly pick them out in winter, when few other warblers are around, and in spring, when the males are in their best plumage. Though pine warblers may sometimes be spotted away from pines, they nest in pines 99 percent of the time and are among the earliest of their group to nest in the spring.

Like other warblers, pine warblers feed mostly on insects. As there are few insects in winter, they may come to your seed feeder on occasion. If you put out a suet feeder, you'll see them all the time.

The trilling warble is similar to but slower than a chipping sparrow's. Just be aware that chippies also sing from pine trees.

Faint streaking
is sometimes visible
on the sides.

The body is olive above and yellow to pale yellow
below, with two white wing bars and split eye rings.

Red-eyed Vireo *Vireo olivaceus*

The red-eyed vireo is one of a group of small, warblerlike birds characterized by a bill that is longer and thicker than that of most warblers and has a little hook at the tip of the upper mandible. Vireos are divided into two clans: those with wing bars and a white or black "mask" around the eyes, and those with no bars or mask, but a white eye stripe instead. The red-eye is in the latter category. Its white eye stripe has a thin black border; above that is a prominent gray cap. For the other characteristics, look for dark olive or greenish back and wings, and white or light underparts. The red eye is not a field mark: it won't be noticeable unless you get a very close look in just the right light.

Like other vireos, the red-eyed feeds almost exclusively on insects. It works its way slowly through the foliage, looking for the caterpillars of moths and butterflies or any other crawly thing. Like all birds, it has keen vision and sees colors well.

This vireo arrives here from its wintering grounds in South America in the spring and either nests here or continues on farther north. In fall, it returns to its winter quarters.

The distinctive nest is suspended between the forks of a branch and secured with spiderweb. The young must all leave the nest at once with the adult; stragglers are left behind. Unfortunately, these young sometimes consist of several

cowbirds, which have parasitized the nest!

The red-eyed vireo has a short, two-phrase song that sounds like *chee-ay*, short pause, *ee-yoo*. It sings this composition over and over, with variations, far longer than most birders want to hear it; but after all, it's not singing for the benefit of birders.

Note the hook at the end of the thick bill and the striping through and above the eye. Most times, as here, the eyes do not appear red.

Chimney Swift *Chaetura pelagica*

Hear a chittering above you in the sky? Stop and look for small, dark birds with cigar-shaped bodies and curved, crescentlike wings. They will be chimney swifts eating on the fly. Their flight is erratic and batlike, with rapid wing beats followed by short glides and frequent changes in direction. They don't land until the end of the day when they gather in flocks, large or small, and descend in a whirl to their roosting site—a chimney or other human-made hollow.

Although you're unlikely to see this little bird in a roosting or perched position unless a fledging falls down your chimney, it is sooty gray or dark brown above and a lighter shade of that color under the chin and on the breast. The very short tail and long, curved wings give it a distinctive "flying wing" appearance. Swallows, which have a similar form, have shorter, uncurved wings.

Only this one species of swift is found east of the Rocky Mountains, and only during the warm months (March to October). For the other five months of the year, the chimney swift gathers in large migratory flocks and heads to the upper Amazon River basin in Peru, Ecuador, Chile, and Brazil.

Some chimney swifts still roost and nest in large, hollow trees, but only in deep forests where there are no chimneys. The nest resembles a

The silhouette above is what we typically see of this species: a small dark bird in the shape of a cigar with two curved wings attached.

half of a cup made from small twigs glued to the side of the wall. The "glue" is a gelatinous substance secreted by the salivary glands.

With the advent of our human developments, these aerobatic birds have increased in number and are now very common over our cities, parks, and countryside. Given all the insects they eat, that's great!

The chimney swift is sooty gray or dark brown on its upperparts and a lighter shade of gray under its chin and on its breast.

House Sparrow *Passer domesticus*

The house sparrow is not a well-liked bird in the United States. Maligned as an exotic import that doesn't belong here, it takes nesting spaces away from our beloved eastern bluebirds and purple martins. It is a scruffy parking lot bird, a dingy city bird, a refuse eater, surely too common for any self-respecting birder to appreciate. Yet those who look with unbiased eyes at the male's striking breeding plumage get a visual treat. The crown is gray, the nape is chestnut, the cheeks are white to gray, and the bib and bill are black. The shoulder is chestnut with one white wing bar. That's a lot of color to crowd into so small a space. The rest of the bird, of course, is just rather sparrowish; ditto for the female and juvenile.

This bird is a weaver finch, not a close relative of our native sparrows. Eight pairs were introduced in Brooklyn in 1850 to help eat cankerworms. That group died out, but later introductions did well enough to spread the species throughout much of the United States.

The breeding male has a black chin,
chestnut wings, and a white wing bar.

You almost never see these sparrows far from human habitation. They nest in any cavity or crevice they can find, often behind some fixture near a building, bringing in quantities of hay and grass, feathers, hair, string, and whatever else they can find. Two broods are usual. They sometimes lay eggs in other birds' nests. These are scrappy and aggressive birds, often stealing from each other and getting into fights. They love dust baths.

Two popular field guides don't even deign to describe the house sparrow's song. Another says it is a long series of monotonous musical chirps. That's true enough, but they're distinctive musical chirps, and the birds sing throughout the year.

House sparrows skittering around on the asphalt of a parking lot as they search for fragments of waste food are a pleasant sight. Look at them closely and think good thoughts. They're among our new "technobirds" and true survivors.

The female and juveniles of both sexes are much plainer than the males.

Tufted Titmouse *Baeolophus bicolor*

A prominent gray crest distinguishes the titmouse from all other birds of its size. Notice, too, the gray upper parts contrasting with the white breast. A buffy patch on its side beneath the wing and stark black eyes complete the description.

The titmouse is one of our most common birds around all neighborhoods, winter and summer. Normally it scours leaves for insects, actively flitting from place to place, even hanging upside down from twigs. But it is a reliable feeder bird and particularly fond of black oil sunflower seeds. Watch it take a seed, fly to a nearby twig, hold the seed with its feet, and peck it open. In less than a minute it will be back for more.

The titmouse is a close cousin of the chickadee, and you'll frequently see two or three titmice in company with several chickadees. In breeding season, however, they are, like most other birds, territorial. Tufted titmice nest in a tree cavity or sometimes in a small birdhouse.

Our bird has a sweet song, much louder than you'd expect for such a small bird. Its *peter peter peter* is a characteristic sound of spring, and it is not above whistling *tweet tweet tweet*. You'll hear squeaky squawks and sucking whistles during food-begging and fighting.

If you are patient, you can gradually train a titmouse to take sunflower seeds from your hand. First, let them learn to eat near you. It helps to have places nearby where the birds can perch while they're deciding whether to brave a possible trap. They may fear your eyes, so squint or close them at first. Be very still. Eventually they may come.

This very common small gray bird is best identified by its crest. If that is not visible, the gray back, lighter belly, and long tail are always apparent. The sexes are alike.

Eastern Bluebird *Sialia sialis*

What species has societies and publications devoted to it and nest boxes put up especially to attract it? The eastern bluebird. Once you see one in the glory of its vibrant blue, you'll be hooked too. The male's back side is an intense sky or cobalt blue. Just as often, though, you'll be glimpsing the rust-colored breast as the bird perches on a wire. Females and juveniles can show much more gray than blue, so be ready to make your judgment by the rusty wash.

The bluebird requires short-grass fields or lawns, in which it goes to the ground to catch its insect prey. For successful breeding, it also needs a very special kind of cavity, not too large or too small, too high or too low. Look for bluebirds on utility lines, fence wires, or the low limbs of trees in open areas. Parents and their young tend to stay together all winter, so any assemblage of three or four small, fat birds on a wire, spaced several feet apart from each other, will usually turn out to be bluebirds.

The bluebird's song is slight and subtle, but sweet—three or four soft gurgling notes.

The male (top) is a glorious blue with a rusty breast, a short tail, and a fat belly. The female (bottom) is less brightly colored than the male but has the same pattern and shape.

Human land-use practices formerly caused a great decline in the bluebird population. More recently, however, thanks to the sustained efforts of many dedicated individuals, bluebirds have made a comeback. Today they are once again a common sight in the open countryside. Making nest boxes and establishing bluebird trails around field edges are very exacting projects. Obtain the proper information before you begin. Bluebirds can be particular, but they will nest in boxes that are appropriately located on a short fence facing a field. Thin metal posts are preferred over trees, perhaps because they deter climbing predators.

In spring, the male brings insects to the female while she incubates the eggs.

Downy Woodpecker *Picoides pubescens*

The downy is our smallest woodpecker. In the suburbs you're sure to see or hear it fairly often. It likes deep woods and dead trees as much as any woodpecker, but can make do with very little of both. You'll see it around your house, pecking food on the smallest of trees or making a home in a dead branch stub that you wouldn't think fit for any bird. It is petite, busy, and mottled in black and white. The male has a spot of red on the back of his head.

This little woodpecker can be mistaken for the red-cockaded woodpecker or the hairy woodpecker, which are also found in coastal Georgia, but the RCW has a large white patch behind the eye rather than the black band of the downy and, being rare and endangered, is very unlikely to be seen in your backyard. The hairy woodpecker (not included here)—also far less common than the downy—is like its cousin in appearance and habit but is noticeably larger (medium size, not small) and has a much larger beak relative to the head size.

Like most woodpeckers, the downy finds its food by scouring the trunks and branches of trees. Preferring middle heights, it seldom comes near the ground, unlike its cousins the flicker, the hairy woodpecker, and the pileated woodpecker. It readily comes to suet feeders, particularly in winter, and in many cases even takes a fragment from the seed tray.

The song of the downy woodpecker is easy to recognize, and you'll begin to hear it often once you learn it. It's a nasal whinny, starting high and then descending, not too loud. There's nothing else like it out there.

A study performed by the Cornell Lab of Ornithology determined the downy to be the most common woodpecker at U.S. bird feeders during the winter. In Georgia, though, the downy usually comes in second behind the red-bellied woodpecker, which can turn up in just about any habitat.

feeder, Oct, 1x · ?

The male (opposite) shows his black-and-white back with a clear white stripe down the middle and a red area on the back of the head. Female (below) lacks the red. Both have small bills and black ear patches.

Eastern Phoebe *Sayornis phoebe*

There are many birds in the group called "flycatchers." In fact, this group includes more bird species than any other group in the world. Most flycatchers are infrequently seen and very difficult to identify visually. One that is likely to be seen regularly by bird-watchers along the coastal area in winter is the eastern phoebe. Another is the eastern wood-pewee (not included here), but that species is here only in the summer.

The phoebe is easy to identify. First, look for a solitary bird on a low perch. Next, observe its black head, white throat, and breast very faintly washed with yellow. The back is dark gray. Last, but important, watch the tail. The phoebe raises and lowers its tail every few seconds; the pewee doesn't.

The phoebe sits on twigs overlooking open spaces, often near a pond or wetland, and darts out to catch insects on the wing, then returns to its perch. It sits low in a bush or small tree, usually no more than 6 or 8 feet above the ground. These birds enjoy the varied habitat of mixed-use farmland, and they love barbed-wire fences, from which they can dart down into the stubble to capture some unlucky grasshopper.

Phoebes like to nest underneath a natural or artificial structure: on a ledge beneath a rock overhang, perhaps, or under your front porch roof or dock.

We remember the phoebe, and have probably been aware of it all our lives, because it sings its old-fashioned name: a buzzy *fee-bee* or *flee-bee*. The pewee sings its name too: a slurred, whistled *peeeo-weee*.

It has the typical hooked flycatcher bill. Occasionally, you can also see some light yellow wash on the belly. Unfortunately, the photos can't show the wagging tail.

The phoebe's dark head and back contrast with its light throat, chest, and belly.

White-throated Sparrow *Zonotrichia albicollis*

The white-throated sparrow is with us only during winter and early spring. Everyone who regularly feeds birds on the ground is apt to see it. This sparrow comes in small flocks of four to eight. It's usually found scratching under your feeder early in the morning and late in the evening, the same time most of the cardinals come. It pecks at small seeds awhile, then whisks away to nearby cover. During the rest of the day it stays mostly out of sight, hiding in thick clumps of honeysuckle or privet along streams, or in tangles by brushy fencerows.

The brown, sparrow-sized body and beak, and the stark white throat are the best identification characters. If you use binoculars, you'll also see a small yellow patch, or "lore," in front of the eyes that becomes bigger and brighter as spring approaches. The back and wings are brown streaked with black. The breast is pale gray with light streaking. This is an interesting species because, like the screech-owl, it is dimorphic: it can appear in two forms. One form has a bright black-and-white-striped head pattern; the other has a black-and-tan-striped head pattern. Ornithologists once thought that these represented male and female, respectively, but we now know that either form can be either sex.

When it is hiding in the brush, we recognize the white-throat by his call note, a short, bright, bell-like sound, but a little raspy, as if the bell were cracked. When dogwoods bloom and native trees leaf out in spring, the little white-throat becomes anxious to get back to its breeding grounds up north. At that time you'll begin to hear the male's song, a plaintive whistled theme, sometimes described as "Old Sam Peabody, Peabody, Peabody." You'll hear that for a few days, a few weeks; then one day, nothing: the bird is gone. You won't see or hear him again until next November.

Adults of both sexes have a clear head pattern, which may be black and white with a bright yellow spot on the lore (top), or brown and tan with a much duller lore spot (bottom). Both photos show the sharply delineated white throat and sparrow bill.

Barn Swallow *Hirundo rustica*

This species is the most widely distributed swallow in the world and breeds all over North America, including Georgia. However, most of them do not spend the winter here. It is included here because it is very common and numerous during its spring and fall migrations.

You can identify these small birds by their blue-black upperparts, rusty neck, and buffy breast. They have long, angular wings and a deeply forked tail—the only one of our swallows with that feature—allowing them to expend less energy in their insect foraging as they sail along. Also, listen for their bubbly, high-pitched chattering and twittering as they snap up another tasty insect. Better check them out the day you see them, as tomorrow they may be continuing on their way south.

Barn swallows migrate in the daytime rather than at night, unlike most migratory birds. Thus, they are quite prominent as they glide smoothly and effortlessly through the air catching all kinds of insects in flight. Over the rest of the United States they are well known and loved for their insect-eating habits, and farmers inadvertently provide perfect nesting sites in barns next to open fields where insects abound.

These aerial acrobats have increased abundantly since humans have provided vertical structures with overhead shelter (barns, wharves, bridges), and they are seldom seen nesting anymore in natural sites (caves, rocky ledges, etc.).

Although there is a fair amount of variability in coloration, the male (below) has a rufous chest and belly, while the female has a lighter belly (opposite). Both share the deeply forked tail.

European Starling *Sturnus vulgaris*

The starling is a bird people rave about, one way or the other. Its rather nondescript black feathers become iridescent green under the right light during the breeding season. In the winter, they are speckled with white. Its strong beak is bright yellow in summer, dark in winter. The wings look very pointed in flight, and when it glides, they appear swept back like those of a jet fighter.

The starling is a bird of ill repute both because it's not native to North America and because of some other bad habits. The species was brought from Europe to New York in the late 19th century and spread rapidly across the country. This aggressive bird, equally at home in city and farmland, often steals the nest cavities of native species such as bluebirds, red-headed woodpeckers, and purple martins.

Starlings are omnivorous: they eat seeds, insects, fruit, and garbage. They can be seen pecking their way across lawns or scavenging in parking lots. They may flock by themselves or with blackbirds. In mixed flocks, they can easily be distinguished by their stubby tail, pointed wings, and less intense black color.

The starling's song is an interesting and not unpleasant assemblage of squeaks and whistles. The imitative myna bird is one of its kin.

Although it is widely disliked because it competes successfully against certain American birds, the starling is as attractive a bird as many others. Let's restrain our value judgments just long enough to see the starling as a wondrous product of nature.

The male and female are alike, and both wear different summer and winter outfits. The nonbreeding, winter plumage (left) is speckled and the bill is dark; the breeding, summer plumage (opposite page) is almost all black and the bill is yellow. In spring and fall, the plumage can be intermediate between the two.

Brown-headed Cowbird *Molothrus ater*

Bird-watchers hate the brown-headed cowbird; scientists find it fascinating. It makes no nest of its own, and instead lays its eggs in the nest of another bird, often a much smaller warbler or other migrant. Its young take over the nest, usurping the energies of the stepparents, sometimes even ousting the legitimate offspring. It thus wreaks havoc on other species. And yet this unusual breeding strategy, similar to that of the European cuckoo, is worthy of study and perhaps even admiration. If only this fascinating bird was rare and beautiful instead of plentiful and not particularly handsome, how differently might we view it.

The mouse-colored female skulks around at dawn, waiting for her target bird to leave its nest. She darts in, quickly lays a speckled egg, then darts away. Sometimes the returning mother detects the new addition and ejects it; other times not. The cowbird's plan works only half the time, but it costs her nothing more than a few eggs.

The male is black or greenish black with a brown head. Its bill is shorter and more sparrowlike than that of other blackbirds, with whom it often mixes. Its song is a short, squeaky whistle/gurgle; the female gives a soft, rattling chortle.

In the past, the cowbird's range was restricted to the vast prairies of the West and Midwest. Like cattle egrets, cowbirds follow grazing animals to catch the insects they stir up. As agriculture and development opened up the woodlands of the eastern United States, the cowbird moved in. Now it is absent only from the deepest unbroken forests, which are few indeed. Thus, rural development is helping to cause the decline of our forest songbirds.

The male (top) has a two-tone color pattern; the female (bottom) is tan or gray.

The male's iridescence, so common in other black
icterids, makes the bird appear glossy blue.

Summer Tanager *Piranga rubra*

This is the South's "summer redbird," so called to distinguish it from the cardinal, which is here year-round and is known simply as the "redbird." Although he is nearly solid red like the cardinal, the male summer tanager is rosy red rather than the fire-engine red of the male cardinal. Also, the tanager lacks a black mask around the beak, which is light-colored rather than orange. The female summer tanager is very different from the female cardinal, having an olive green back and yellowish underparts. Overall, the female is rather plain, and the light-colored bill does not stand out as the bright orange bill of the cardinal does. Distinguishing the summer tanager female from the female scarlet tanager, on the other hand, is very difficult. The major method of separating them is geographic: the scarlet tanager nests only in the northernmost part of Georgia or farther north, and is seen on the coast only during migration (usually April and May).

The summer tanager nests in deciduous hardwoods such as oaks and hickories, but also nests in pinewoods. Except during migration, it is seldom seen in the cities and suburbs (unlike the cardinal) unless there are lots of trees and water nearby.

This tanager forages mainly for insects in the treetops, and while not really shy, is hard to see among the leaves. However, its call note can easily identify it. Think of an old-fashioned metal frog clicker's *tick-tock*, with perhaps a little more of a *tick-cha* than a *tock* at the end, and often with more syllables added as *tick-tock-tuck-tuck*. The song is similar to a robin's but less energetic, and is also like the scarlet tanager's song without the fullness. Given these difficult distinctions, just listen for the *tick-cha* that is commonly given.

The immature first-year, and sometimes second-year, males may have various degrees of rosy red mixed with olive green as they slowly change from the juvenile's female-like plumage to that of an adult, as seen here (left).

The male (opposite top) has a beautiful rosy red color and no black mask, unlike the male cardinal. The female (opposite bottom) has a yellow-green color pattern completely different from the female cardinal.

Eastern Kingbird *Tyrannus tyrannus*

Although it is widely known in the countryside as the "bee martin," the kingbird will eat any insect it can capture and swallow. This conspicuous flycatcher is the only widespread kingbird species in the eastern United States. You can easily spot one sitting in a tree in an open field or pasture, on a roadside power line, or on any other open perch from which it can dart out to nab a passing insect or watch for these tidbits.

You might at first mistake a kingbird for an eastern phoebe, but the kingbird's larger size, relatively larger beak with a hooked end, and the very noticeable white border on the tail quickly distinguish it. Like the phoebe, however, it has black or gray upperparts and white underparts. It rarely displays its concealed red crown patch except during mating season.

The eastern kingbird is a common summer resident from March into September where open areas afford easy access to insects, its main diet, spiced up occasionally with wild fruit and berries. It is an aggressive and fearless defender of its nesting territory, relentlessly attacking crows and hawks, and other kingbirds, to drive them away. At summer's end, however, kingbirds congregate together to migrate in day-flying flocks to South America for the winter.

Unlike the phoebe or the similar wood-pewee, the kingbird is generally seen rather than heard. When it does vocalize, it has a sharp, shrill chattering cry with a swift upward flutter at the end.

The breast is a pure white, contrasting sharply with the dark
upperparts. Note the distinctive white-tipped tail.

Cedar Waxwing *Bombycilla cedrorum*

The sleek cedar waxwing, with its sharp crest and black mask, is among our most elegant birds. Brownish amber overall, its tail is tipped with a band of yellow. You may see red waxy tips on some secondary wing feathers if you get a close look at an adult.

Waxwings come to berry bushes in large, bubbly flocks, seeming to pounce on the bush from the sky. They strip it of berries and then move on, often leaving quite a mess beneath.

Flocking is what waxwings do best. Indeed, they are seldom seen singly except during breeding season, which comes late in the year after berries have ripened. Flocks are usually tight and well controlled, wheeling this way and that with precision. In flight they could resemble starlings but are smaller and browner.

Berries are waxwings' staple food, but they also eat insects. Their strangest food, however, is flower petals—apple, for example, and tulip tree. In courting rituals, the male and female pass these back and forth to each other. They nest in the northern Piedmont and the mountains.

The waxwing's song is a high, thin, bell-like trill, like the ringing of tiny chimes. You hear it not as an individual song but as the music of an orchestra.

loved fermented pears in back yard, in ye Merry crowd, never saw sober.

The subtle breast colors, yellow-tipped tail, black face mask, and head crest are present in both adults and juveniles, but adults (opposite page) also have red-tipped wings, which the first-year bird (above) lacks.

Purple Martin *Progne subis*

The purple martin is our largest member of the swallow family. Like all swallows, it has very long wings; short legs; and a short, flat, triangular bill that is designed for catching insects on the fly. All swallows have a notched or forked tail, with notched being the case for the martin.

The male is a uniform, glossy steel blue, or just dark if viewed only briefly in flight. The female is much duller, to the point of being brownish, and much less uniform; her underparts have a light sooty color. The male and female are thus sexually dimorphic, unlike most of the other swallow species.

Martins are among the earliest migrants to appear in Georgia, arriving in late February or early March, and departing from early July through August to return to the Amazon basin for the winter. Prior to leaving, they gather in huge flocks to roost and then all leave together as a "kettle."

The purple martin's "song" is a series of early morning chirps, chattering, gurgles, and clicks that have a musical quality that is difficult to describe in words.

Purple martins have long been a friend to humans. Native Americans hung up gourds to attract them to nest nearby so that they would eat pesky flying insects, and European colonists followed their example. This practice of providing nest gourds (and the multilevel bird condominiums also used today) seems to have caused martins largely to stop nesting in old woodpecker holes or other hollow cavities. Artificial nesting homes are now common sights in the countryside, where martins are much appreciated for their insect diet; nevertheless, there seems to be little truth to the rumor that each martin eats 2,000 mosquitoes daily.

The male is a glossy black (opposite top); the female is brown and sooty (opposite bottom). Artificial nesting homes are now common sights in the countryside, where martins are much appreciated for their insect diet.

Great Crested Flycatcher *Myiarchus crinitus*

INCHES

No doubt about it, this flycatcher is one bird you won't miss if it's around. Its harsh *wheep*, given about every 30–60 seconds on average, can be heard for quite a distance. Although it doesn't really hide, the great crested can be hard to spot in the leafy deciduous trees in which it likes to perch. If you give it a minute, however, it will fly out to catch an insect or will let out another call you can use to home in.

You'll see a cardinal-sized bird with long, rusty tail feathers; a gray head and neck; olive or brownish back; and a nice mellow yellow under the breast area. The bill is a typical flycatcher's: medium length and quite wide. The wings are edged with cinnamon rust and have two white wing bars, making this species one of the more colorful flycatchers.

The great-crested nests in tree cavities or nest boxes, preferably near open spaces such as fields, pastures, or orchards. One of its more interesting habits is to add a shed snakeskin to its nest; no one knows why. Pairs produce one brood per year. This flycatcher is very territorial, driving off others of its species fiercely, which pretty much makes it solitary during its summer stay here in the South. Oddly, however, it ignores the crows and hawks that snatch eggs and young out of its nests.

Come fall, the great-crested retreats back to the tropics or southern Florida for the winter, flying at night like many other migratory species.

BACKYARD

Male and female are alike with gray on the throat, yellow on the belly, and cinnamon wing edges. Note the distinctive head crest (opposite).

Eastern Towhee *Pipilo erythrophthalmus*

Our towhee is a ground bird kin to the sparrows. It scratches around among the leaves looking for seeds and insects. The male has a striking black head and back, a white belly, and rufous sides. The female is the same, except her upperparts are brown instead of black. The eyes of both sexes are red-orange.

The towhee continually flicks its tail, spreading it open to show the two hidden white feathers. This white-spot flicking is just as much a part of the physical appearance of the bird as any other feature, so be alert for it.

Towhees will come to bird mix spread on bare ground, and sometimes fly up to a feeder. Like brown thrashers, the only time they seek height is when they're singing in the spring. The nest, located near the ground, is well concealed and difficult to find.

Here's another bird whose call sounds very much like its own name: *toe-weee*. The male's spring song is *drink-your-tea-eee*, but often it says only part of that, and sometimes the *tea* is a whistle. A pair of towhees feeding in the same vicin-ity will keep in con-tact with each other by making frequent *chip*s and calls.

The male towhee (above and top left) is most often seen in his familiar ground habitat. The female (opposite bottom) has brown where the male has black.

Red-headed Woodpecker *Melanerpes erythrocephalus*

The red-headed woodpecker is recognized by its black-and-white wing pattern and dark red head. No other woodpecker has an entirely red head, yet people often mistakenly give its name to the red-bellied woodpecker, whose crown is the color of an orange-red Crayola rather than dark red. The juvenile red-headed that you see in midsummer is different: its head is gray-brown, and its wings are not quite as black.

Like most other woodpeckers, the red-headed searches trees for insects hiding

in or under the bark. Or it may dig for a wood-boring grub. Working hard and taking turns, the male and female can hollow out their nest hole in a matter of days. They prefer to build high in a dead pine, often in a swampy area. In the city, they like large hardwoods growing not too close together, such as in a park or along a street. They would probably use our dead pines if we left a few standing.

In winter, this woodpecker will take sunflower seeds from a tray feeder or go to suet on the side of a tree. On the tray, it is feisty and selfish, often attacking other birds with its spearlike beak.

The red-headed woodpecker sings loudest in spring, when rivalries between males give rise to raucous singing contests. But it's not much of a song: a rasping whir or a barked nasal squawk. The red-headed will drum loudly on a tree to stake out its territory.

This species is declining throughout its range, possibly because starlings are taking over its nesting cavities.

(my pet favorite)

The solid red head and contrasting black-and-white wings show equally well whether the red-headed is on a tree or in flight. The male and female are alike.

Gray Catbird *Dumetella carolinensis*

The catbird is best seen in relation to our other two mimids: the brown thrasher and the northern mockingbird (both in this book). They're called mimids because they mimic other birds as well as various noises. They have longish tails that they tend to cock upward, and they have businesslike bills. All three species like the same foods: insects and berries.

In color, these mimids are very different. The large brown thrasher is a beautiful chestnut brown with heavy breast streaking. The mocker is gray with black and white on the wings (seen especially in flight). And the catbird is all gray—dark gray. Its decoration is inconspicuous: a black cap and a patch of dull red under the tail.

The catbird's song is a slurred and garbled (but pretty) mishmash of single notes. The thrasher sings in pairs of notes (mostly) and the mocker sings in triplets (often). But the catbird's most distinctive noise—from which it gets its name—is *meeow,* its alarm call. You'll more often hear catbirds than see them, for they love to stay hidden in dense tangles of bushes and small trees. Look for them where their favorite food is: grape arbors and, in fall, around pokeberry bushes, on whose purple berries they gorge.

Coastal Georgia is near the southernmost part of the catbird's year-round range. In summer, many go farther north and in winter, farther south, but some are always here.

Note the solid gray of this species, broken up by a black cap; note also the difficult-to-see rufous undertail coverts (left).

Northern Cardinal *Cardinalis cardinalis*

"Redbirds," we call them in the South, because of their nearly unmistakable brilliant red plumage and red-orange beak. So common are they that we take them entirely for granted; yet people travel thousands of miles to see these striking red beauties. The sexes are different. The male is almost completely red except for a grayish wash over the wings. The female has a more brownish or grayish coloration. Juveniles of both sexes look like the female, but have a brown, not red-orange, beak.

The heavy mandibles show immediately that the cardinal is adapted to crushing seeds, not spearing insects. That stout beak can draw blood from a human's finger. Watch as the bird goes through a pile of sunflower seeds. Seed husks drop out of the sides of the beak as the mandibles grind back and forth to extract the inner meat.

Cardinals will turn up anywhere and are certainly adapted to our suburbs, but they seem to like swampy woods best. In winter, when they aren't being territorial, flocks of 30 or more may take advantage of feeding stations. The range of the cardinal has expanded northward as bird lovers have provided more food during winter.

Cardinals sing mostly in spring, when you'll hear their pleasant *kyeeer, kyeeer* or *purty, purty, purty*. At other times, they utter a single metallic *chip* that you can easily learn to recognize.

Harry Kressing's novel *The Cook* describes redbird hunts with gourmet feasts following. Hunting for the millinery trade once decimated bird populations. Fortunately, it is now against the law to kill or harass any songbird, or even to possess a feather of one.

The prominent crest, orange bill, and black face make this male highly visible as he guards his territory (top). The female (bottom) is brownish gray with some red on the crest, wings, and tail.

Yellow-bellied Sapsucker *Sphyrapicus varius*

Although this name may be an insult when thrown at another person, it is quite appropriate for this woodpecker with its yellowish underparts. The yellow-bellied also has the plump body and stout, pointed beak typical of woodpeckers, along with a red patch on a part of its head—in this case, the forehead and forecrown. The male has a red throat patch; the female differs by having a white throat patch. The most distinctive identification feature is a broad white wing patch running from the shoulder almost halfway down the wing; it can be picked out from a distance.

The yellow-bellied sapsucker migrates the longest distance of any eastern woodpecker. It spends its winters in the Southeast, and then heads back to the northern United States and Canada to breed during the spring and summer. There is essentially no overlap in its summer and winter ranges.

The sapsucker may move around considerably from area to area during the winter, or one may pick a tree it particularly likes and peck row upon row of regularly spaced, fairly deep holes in the trunk. It then uses its brushlike tongue to lick out the sap that flows into the holes—thus its name. Ants and other insects that come to the sap are included in the diet as well. The sapsucker is

The male (above) has a red throat; the female (left) has a
white throat. Both sexes have white wing patches, which help
to identify this species; yellowish underparts; and a red cap.

our only woodpecker whose diet consists of more plant material (sap) than
insects and other invertebrates. Unfortunately, damage and even death to the
tree can result when the sapsucker makes hundreds of holes in it.

This woodpecker is generally quiet in the winter, but it can produce a cat-
like mew or whine (sounding like a downward-slurred *cheeerrrrr*) that is also
distinct and identifies it unseen. It tends to be shy and hides behind the tree
trunk when humans are around.

Eastern Screech-Owl *Megascops asio*

The eastern screech-owl is a small, chunky bird, no more than 8 inches tall but with a wingspan of nearly 2 feet. Most individuals are rusty reddish brown with white markings on the wings and chest. Some are gray, however, and a rare few are drab brown. Different color phases, or "morphs," like this occur in several species of birds. The causes are not yet clear.

In our area, screech-owls show up wherever there are trees with cavities. Common though these birds are, they conceal themselves well. You're more likely to hear one than to see it. Most of their singing is done in spring and summer. One song is a descending trill or whinny; another is a low, monotonous rattle, very soft. When protecting their young from intruders, the adults make strange barks and popping sounds.

Screech-owls eat birds, small rodents, and large insects. They attack silently on wings muffled by fuzzy feather edges, making no more noise than a butterfly.

Many people render screech-owls more observable by placing nest boxes in their yards. A good box might be 8 inches square and 12 inches tall, with a 3-inch hole near the top. Owls don't want a perch and don't need nest material. Place the box 12 feet above the ground, or as high as your ladder will reach.

Owls occupy the boxes as early as December. In late afternoon, or even in midday, the owls may bask in the box opening where you can get a good look at them. The young—two or three—will fledge in May or June. The chicks have gray, downy feathers. Sometimes during the day, songbirds will discover a screech-owl roosting in a thicket or under a canopy of vines and "mob" it, fussing loudly and hopping around excitedly. The owl usually endures this nuisance stoically.

Most eastern screech-owls are rusty brown (top left). The gray phase (bottom left) is less common in the South; the owl shown here has its "ears" flattened. The other two (top left and below) have them raised. Note the yellow irises.

Northern Bobwhite *Colinus virginianus*

The male's ascending *bob-white!* whistle gave this bird its name. In the South it is also known as a "partridge." Before humans took over much of the countryside with our suburbs, the bobwhite was one of our best known birds. It is more often heard than seen, but it will come to eat seed under a feeder if thick undergrowth is nearby.

The bobwhite is quite beautiful with its varying shades of brown, gold, gray, black, and white. The male's head has a black stripe through the eye flanked by white stripes; the female's head is similar, but the white is replaced with buff and the eye stripe is brown.

Bobwhites tend to stay on the ground unless forced to fly when danger approaches. In winter, typically 10–30 birds will congregate in a covey. They roost together for safety and warmth at night, forming a circle with their tails inward. Just before you step on it, the covey will explode in flight in all directions.

The bobwhite is the only native quail in the East. It is a permanent resident of open areas near thick underbrush—on farms, pastures, grasslands, and fields. Because they do not migrate even from northern regions, many bobwhites die when a harsh winter occurs. In the summer, when the covey disperses to breed, each pair on its own territory feeds on insects, berries, and roots. In winter, they eat leftover crop seeds such as corn along with the seeds of weeds.

A male bobwhite (bottom) warily looks around with head held high, showing the nice brown-and-white facial pattern along with the beautifully streaked chest, flanks, and back. The female's pattern (top) is similar to the male's, but she has browns and tans on her head.

Eastern Meadowlark *Sturnella magna*

The beauty of the eastern meadowlark is enough to make your heart melt. Fortunately for us they are plentiful, but mostly in the open country during the winter. It's worth taking a winter tour of our outlying farmlands just to catch sight and sound of them.

Before you ever see them, their song will win you over: it's a clear, whistled *see you-see-yeeer*, with the last part down-slurred. So sweet, so clarifying, so uplifting. They also have a call, a guttural noise you wouldn't associate with them at all: *dzrrt*, with some chatter. Listen for that too in the fields, and realize it's coming from the same bird.

Meadowlarks are usually seen in small flocks of 5–15. When they take flight, you can recognize them by the white feathers on either side of the tail. When they glide in for a landing in a field, they look like fighter jets with down-slanted wings. Perched on a fencepost or wire, facing you, there's no mistaking the bright yellow breast with a black v. When facing the other way, however, the meadowlark shows only brown. Notice the large, pointed beak, which puts the meadowlark in the blackbird family, kin to redwings and grackles.

Eastern and western meadowlarks are an example of a species that has recently split into two from a parent species. Outwardly they are difficult to distinguish; their voice, however, tells them apart. The birds themselves know their differences and do not interbreed, even when their ranges overlap in Texas and the Southwest.

Meadowlarks build their nest right on the ground, sometimes interlacing growing grass over it as a roof. The average territory for a nesting pair is 7 acres. Farm equipment destroys many nests, but fallow fields and edges allow the species to persist despite human activities. Both adults and young eat insects of all kinds, and add waste seeds and grain to the diet in autumn and winter.

Eastern meadowlarks have a brown back, long bill, and yellow-striped head. The black v on the chest is partially visible (opposite). Note how well the bird blends into its background.

American Robin *Turdus migratorius*

When robin redbreast sings his best, there can be no doubt that springtime has arrived. Unfortunately, they leave the coast for the summer to nest in more northern areas. There is no finer song than their springtime *cheerily cheer-up cheerio*, nor a better sight than robins hustling in the grass for worms.

Most people recognize robins: tawny red breast, dark back with black head. The color of the females is a little duller than that of the males. Juveniles sport a speckled breast in summer and have an endearing gawky quality.

You can chance upon robins almost anywhere: in fields and yards, in orchards

Worms are a major diet source for robins.

and woods. They usually stay low because they feed on the ground, eating worms and bugs. Don't expect them to be interested in the seed at your feeder.

The nest is made of twigs and straw, and sometimes bits of plastic, bonded with mud or clay, set out on a branch 10 or 20 feet above the ground. The eggs are a beautiful sky azure. Robins often produce two or more broods per season.

Robins are found in flocks both large and small. A flock observed during a Christmas count was tallied at 1 million birds! But more often you see them in groups of 5–20 birds. When they're flying overhead, you can identify them by their spacing: they seem to like staying 10–15 feet apart.

When you see a robin cock its head at the ground, run a little way, stop, and cock it again, what is it doing? Listening? A scientist who decided to figure it out lay down on the ground to find out what she could hear. She didn't hear much, but she discovered that she could see a lot of insects. And so, probably, does the robin.

Hangs about in flocks sometimes, usually not but
sdo's. all outa.

The male robin (top), here shown in spring plumage, has a black head and white eye ring as well as the typical orange breast. The female (bottom) looks similar, but her colors are more subdued.

Red-bellied Woodpecker *Melanerpes carolinus*

The red-bellied is such a common woodpecker in our area that you've surely seen or heard one in your backyard. The head of the male is bright orange-red. At your first meeting with this bird you might have thought you were seeing the red-HEADED woodpecker; that bird has a very dark red head, however, with striking black-and-white wings and body. From a distance, our fellow looks grayish because our eye blends its fine black-and-white mottling. Some people call red-bellied woodpeckers "ladder-backs" because of that pattern. A closer look will reveal the mottling and will also show that the whole head isn't orange-red, just the back and top of it. The cheeks are the same light buffy gray as the belly. The female has even less red, just the back part of her head.

The red-bellied woodpecker willingly eats suet and takes sunflower seeds from a feeder, sometimes fighting for sole possession. It opens seeds by placing them in a crack of tree bark and pecking.

The red-bellied's habits are those of most woodpeckers. It digs its nest hole in a dead tree snag and lays eggs that are round and white, like those of most cavity dwellers. It spends its time on the trunks of trees searching for insects. In spring, it stakes out a territory by drumming on whatever makes a loud noise, including gutters and sides of houses.

This woodpecker's call is a rattle or chortle, intoned in various ways. Distinguishing it from the call of the red-headed takes a little practice. Sometimes you must listen twice to be sure you're not hearing a kingfisher or a great crested flycatcher.

The flight of most woodpeckers is described as "undulating," because they flap in spurts. Between spurts you can see the body arcing like a dart, wings folded. Once you know how woodpeckers fly, you can spot one a mile away.

This species has red on the crown and nape. The red covers the whole crown of the male's head (right) but stops at the top of the crown for the female (left).

"Our" woodpecker. Atlanta
(Mistakenly called flicker in past)

American Kestrel *Falco sparverius*

Our smallest bird of prey is a member of the falcon family. While similar to the peregrine falcon in having a dark facial pattern, the kestrel is only about half the peregrine's size and is more richly colored. The kestrel has two facial stripes: one passing from the crown to below the eye, the other on the side of the face connected at the cap and running to the top of the neck. The male has a blue-gray forehead, cap, and wings to go with a rufous chest covered with small black spots and a rufous back. The female lacks the blue-gray except on the cap, but retains the varied brown to rufous on the back and chest areas. No other falcon has a rufous back. As with all falcons, the tail is disproportionately long compared with the body; it is also rufous.

You typically see this bird sitting on a telephone line along a country road, looking for large grasshoppers or small rodents in the fields or grass. It can also be recognized by its ability to hover in one spot while it views the ground beneath it. In this respect it is like the osprey, which also hovers over its intended prey, but over the water instead of the ground. Because of its size, the kestrel is also known as the "sparrow hawk," but it seldom feeds on birds.

A few kestrels can be found along the coast during summer, but they are much more frequent in the winter, when those that summer in Canada and the northern United States migrate south. Some of them go all the way to South America, but many more winter in the southern states.

Unlike larger hawks, kestrels nest in tree cavities and even can be enticed into artificial nest boxes when there are no convenient trees around.

The kestrel's call is a repetitious *killy killy killy killy*.

The male kestrel (left) displays a beautiful rust color on his back and tail, gray-blue on his wings and crown, and a black-and-white pattern on his face. The female (opposite) has a similar color pattern, but with no blue-gray on her wings.

Northern Mockingbird *Mimus polyglottos*

This is the state bird of Florida, and it is common throughout the South. It is "northern" only to South America!

The mockingbird is generally gray with accents of black and white. It has a habit of flexing and extending its wings, showing their large white patches. You see these too when it flies, along with the white edges on the tail. A sleek gray bird with white wing flashes is bound to be a mocker.

Of all the suburban avian singers, only the brown thrasher can rival the mockingbird. The mockingbird's own specific song is long and complicated. But it also mimics other birds and sounds from the surrounding environment.

Sometimes you truly do not know whether you're hearing another bird or the mocker imitating it. And does it sing! Endlessly, for hours, even at night, particularly in the spring. The call, surprisingly, is a harsh *tchack*. The mockingbird will repeat phrases of its song many times, often in groups of threes. Its cousin the brown thrasher

The flashing white wing patches show best when the bird is flying or hunting insects.

sings in doublets. Its other cousin, the darker gray catbird, sings a garbled song of single notes with little repetition. These three species together are our "mimids," so-called for their mimicking abilities.

The nest is within 10 feet of the ground in a tree or bush, and the mockingbirds will attack anything that comes near it.

While some birds get pushed aside by the advance of civilization, the mockingbird thrives on it. The more cell phone towers the better. The more little starter trees on a lawn of manicured grass, better still. House cats? A mere distraction. Learn to appreciate this very special bird, for it seems destined to survive every obstacle we throw at it.

Note the overall gray color with a lighter chest. The body is sleek with a long tail.

Brown Thrasher *Toxostoma rufum*

Our Georgia state bird belongs to the same family as that great imitator, the mockingbird. The brown thrasher, however, seldom copies other birds' songs and is considered to be the best vocalist in the United States. Its own song, performed in spring from a perch 20–30 feet above the ground, is a series of notes usually sung in pairs. From low in the bushes, especially in the evening and early morning, it will also utter a sharp *chuck* or shushing hiss.

At a distance, a thrasher darting across a shadowy country road could be confused with a female cardinal. But the thrasher is rustier, larger, and has a longer tail, which is often pointed smartly upward when the bird is perched. Notice also the boldly streaked breast, long curved beak, white wing bars, and pale yellow eyes. Another brown bird with a speckled breast is the wood thrush (not included here), which migrates through our area in spring and fall, often staying the summer; that bird, however, has brown (not yellow) eyes, a white eye ring, and a much shorter tail.

Except in spring, when the males sing from perches, brown thrashers are ground birds. They scratch in the leaves, shuffling and overturning them with gusto in their search for food. They can run quite fast along the ground. Insects and fruit make up most of their diet. They are not common at bird feeders, but may occasionally take some seed from the ground.

The egg is a lovely mottled thing, but it takes luck and skill to find the thrasher's clutch hidden low in the bushes. Of course, if a pair nests in the hedge by your mailbox, it's yours for the viewing.

One of John James Audubon's most dramatic paintings shows four brown thrashers ganging up to protect a nest of eggs from a marauding black rat snake. Nowadays thrashers are much more likely to be destroyed by house cats running loose in the neighborhood than by snakes.

Both sexes have rusty brown wings and back; a streaked breast; yellow eyes; a long, uptilted tail; and a beak meant for business.

Brown thrashers spend a lot of time on the ground
hunting insects in the grass and in leaf piles.

Blue Jay *Cyanocitta cristata*

Jays of various kinds are found throughout North America, but the only one in our area is the blue jay. It belongs to the corvid family, which includes crows and magpies. All corvids are smart, noisy, mischievous birds that tend to be scavengers and opportunists. A blue jay will eat anything, from birdseed at a feeder to eggs or babies in a nest.

The blue jay averages 10 inches in length. Its head crest is usually not distinctive. Although we tend to take this common bird for granted, seen with new eyes its subtle colors are remarkable: mauve, blue-gray, blue, and turquoise. The black-and-white markings are striking. The underside is gray, turning to white on the undertail coverts at the base of the tail.

Our jay likes to build its nest at a fork near the trunk of a tree. In early spring, it may construct one or two practice nests before starting on the real one. It likes to build with wet leaves but will use other materials too. It behaves very quietly around its nest, hopping slowly toward it from branch to branch in order to fool predators. It needn't fool you, though: just watch for jays carrying nest material.

Jays are famous for squawking loudly at cats and snakes, but their favorite targets are crows, hawks, and especially owls. A flock of 8 or 10 jays can make a great horned owl really nervous as they fly around and past it, sometimes actually striking the big bird's head and feathers. When you hear mobbing jays giving their most raucous, most frenzied calls, go see if there isn't an owl or hawk nearby.

The blue jay has a short, sweet, flutelike song, and a throaty rattle when it isn't happy.

Blue jays have an unmistakable pattern of blues, blacks, and whites. The blue colors can vary from bright blue to purple-gray depending on the lighting conditions. The blues are the result of refraction of light and not pigmentation.

Mourning Dove *Zenaida macroura*

One bird-watcher in the Southeast who each day listed the birds he saw claimed to have seen a mourning dove every single day of the year. Are they really that common? Not necessarily, but you have surely seen one.

The mourning dove has a tan body with brown and gray wings and gray tail, and here and there a spot of black. Under good lighting conditions you may see some iridescence, especially around the neck and head, which seems small in comparison with the body. The slender beak is slightly down-turned. Strange and special: the wings squeak when the bird flies! The flight is swift and direct, with wings beating quickly and the long, pointed tail trailing behind.

On the ground, the bird walks methodically in the open, pecking at seeds, moving its head back and forth rather like a chicken. It appears to be unable or unwilling to scratch the ground or turn over leaves.

Don't think of the dove as necessarily a bird of peace, for on the feeder it will raise its wings in threat and chase off other birds, including its own kind.

Both sexes are light brownish gray with scattered black spots and a long tail that looks very pointed in flight.

The mournful cooing of doves is a characteristic sound of spring and summer. It issues from low, dense shrubs or from perches high in the pines, where the birds build their flat, flimsy nests. Unlike the many species that feed their young on insects, doves nourish their chicks by regurgitation: poking their head right down the chick's gullet to deliver a special food called "crop milk." Adults eat small seeds, even the minute seeds of violets. Their beak clearly lacks the crushing power of finches, cardinals, and sparrows.

Mourning doves are classed as game birds, suitable for sport hunting. This puts them in a class with turkeys, quail, pheasants, and ducks. No wonder the doves we see at our city feeders are very wary: in south Georgia, they are fair game for hunters, and are also preyed on by hawks, owls, and house cats.

The head seems too small for the round body and long neck. The bill is slim and down-curved.

Northern Flicker *Colaptes auratus*

What's that largish brown bird with the white rump that flies up from the ground when you approach? It's a northern flicker, a member of the woodpecker family. Until recently we called it the "yellow-shafted" flicker, for it shows a flash of bright yellow on the underside of its wings. A similar bird in the West was called the "red-shafted." Zoologists decided these were the same species, so they renamed both birds (along with the gilded flicker) the northern flicker.

Our flicker has a black bib and a red swatch on the back of its head. The belly is strongly spotted with black. The male has a black moustache; the female doesn't. If you see several together on a tree trunk during mating season, check for moustaches to determine each bird's sex. Then you'll know what sort of interaction you're witnessing—love or war.

Flickers are common in our suburbs. Elsewhere they prefer the edges of swamps or open areas with just a few trees. Like other woodpeckers, they ordinarily nest in cavities, usually drilled out by themselves.

The flicker's long tongue is good for lapping up ants—both workers and grubs—but it eats other insects too. Berries, including poison ivy, also make up a large portion of the diet.

In the breeding season, you can hear flickers singing *wicker-wicker-wicker* or one of many variations on this. The bird advertises its territory with a short, sharp, screaming cackle. Flickers also have a one-note call, a harsh *screep*, used year-round.

In earlier times flickers were hunted, for they are good to eat. Nowadays their greatest enemies are owls, hawks, cats, and cars.

The male sports a black mustache. Flickers are frequently seen on the ground looking for insects.

The female does not have a black mustache. Both males and females have a black breast band, a speckled breast, and a red dash on the back of the head.

Common Grackle *Quiscalus quiscula*

A medium-sized black bird without a red shoulder patch is most likely a common grackle. Sometimes called "purple grackle," it may show a bronze, purplish, bluish, or greenish sheen under various lighting conditions. Admire the color, but don't use it as a guide to identification. Instead, look for the long, slightly keeled tail that widens toward the end. The beak is large, and the eyes are pale yellow.

The common grackle is smaller than its cousin the boat-tailed grackle, and much quieter and better behaved as well. Red-winged blackbirds are considerably smaller and have a crescent of red or orange or white on their shoulders. Starlings are short and chunky, with hardly any tail at all, and you can sometimes see their pretty speckling. Cowbirds are much smaller than grackles and black with a brown head (males) or mousy gray (females).

Grackles often come in huge flocks that descend on your lawn. Listen to their noisy squeaking and creaking as they cover the ground, and watch them turn over leaves in search of nuts and seeds. They especially prefer small acorns such as those of the water oak. In fact, their upper mandible is equipped with a special ridge that helps them crack the nuts open. Grackles strut about as if they think they own the place, but they startle easily. The whole flock may take flight at once just because you raised your window shade or opened the back door.

The grackle's call, according to some observers, is *koguba-leek*.

In winter, grackles cross the sky in huge flocks, often in long, wavy lines. Other blackbirds may be mixed with them, including red-wings, starlings, and cowbirds—all roosting together. Forget the saying "birds of a feather flock together"! Look to see what's there.

The grackle may appear bluish black (bottom) or mostly black (top), depending on the lighting conditions, but it always has a long, wide tail and pale eyes.

Rock Pigeon *Columba livia*

Field guides list four species of pigeon (*Columba*) for the United States: band-tailed, red-billed, white-crowned, and rock. The first three are highly restricted in range, but the rock pigeon is found in nearly every town. We know it simply as a "pigeon." All the pigeons in the United States are descended directly from imported domesticated birds.

Pigeons are excellent fliers and are quite beautiful as they wheel and turn in flocks. In some cities, people raise them in rooftop cages and have contests and "wars" with one another's flocks. Charles Darwin is among the most famous historical pigeon fanciers. "Believing that it is always best to study some special group," he wrote in chapter 1 of *Origin of Species*, "I have, after deliberation, taken up domestic pigeons" (1859; Penguin reprint, 1977). There are many different breeds of pigeon, some of them very strange: carrier, short-faced tumbler, runt, barb, pouter, turbit, Jacobin, trumpeter, laugher, and fantail, to name the most important.

Rock pigeons can be a mixture of colors ranging from pure black to pure white, but the most common pattern is dark with an iridescent neck (left) or with a light gray back and wings striped with two dark wing bars (above).

Rock pigeons come in many colors, including white, reddish brown, and black, with gray-blue being the most common; and there are mixtures. But when left to breed on their own, the birds have a tendency to revert to the ancestral form. Let Darwin describe that form: "I crossed some uniformly white fantails with some uniformly black barbs, and they produced mottled brown and black birds; these I again crossed together, and the one grandchild of the pure white fantail and pure black barb was of as beautiful a blue color, with the white rump, double black wing-bar, and barred and white-edged tail-feathers, as any wild rock-pigeon!"

Pigeons inhabit our cities because the stone and concrete buildings are like the rocks on which the ancestral wild species lived. Instead of despising or ignoring our rock pigeons, let's admire their iridescent colors and enjoy their swift flight and aerial acrobatics.

Eurasian Collared-Dove *Streptopelia decaocto*

The collared-dove is clearly larger and much paler than our well-known mourning dove. Furthermore, it has a distinct black collar on the nape from which its name is obviously derived. This species is new to the United States but is spreading rapidly throughout the country from its initial start in Florida.

During the early 1970s, a few dozen European collared-doves were brought to the Bahamas to establish a local breeding colony. Prior to this, the collared-dove had spread rapidly and widely over Europe from India and the surrounding area. A burglary in the Bahamas allowed many to escape, and then the rest were set free. They established a thriving colony and quickly expanded their range to reach the southern part of Florida in the 1980s. Since then they have continued to expand and now can be found in many states. So far, they have not caused any problems for our mourning doves.

Unfortunately for true birders, who like to be positive about which species they are seeing, the Eurasian collared-dove has a very close look-alike called the African collared-dove (*Streptopelia roseogrisea*). The African collared-dove is found in the United States only in a few large cities, though, where its survival depends on human handouts. The only way to distinguish the two species is by their different voices and by examining the pattern of the underparts of the tail.

As African collared-doves are occasionally released during weddings or other occasions, you may find it difficult to be sure your bird is a true wild species—that is, a Eurasian collared-dove. Because the Eurasian collared-dove can produce as many as six broods a year, however, you may not have to wait until spring to hear its frequent trisyllabic *kuk-kooooooo-kook*, which will immediately confirm its identity.

Note the larger size and lighter color of this dove when compared with a mourning dove. The black band on the back of the neck is a distinct identification mark.

Pileated Woodpecker *Dryocopus pileatus*

The pileated woodpecker is a magnificent reminder of what its recently extinct cousin, the even larger ivory-billed woodpecker, looked like. Early woodsmen, with good reason, dubbed the pileated the "gawd-almighty" bird. Big as a crow and with almost as much black, it glides into your woodlot like a dream or vision. It may peck at a hole it wants for a nest. It may course the trunk of a tree, up and down, looking for ants or other insects. More likely, it will fly right down to the ground to tear apart a rotten log, using its large, chisel-like beak to flake off huge chunks of wood. Grubs and beetles it will take, and centipedes, millipedes, and spiders; but what it really craves are large, black carpenter ants. "Log cock" is one of its local names.

The model for the cartoon character Woody Woodpecker, the pileated gets its name from its conspicuous red crest, or "pileus." Its loud, crazy cackle—a series of 10–15 *kuks*—rings through the woods. The feeding peck (as distinct from territorial drumming) is loud and slow with no rhythm. The bird's flight pattern as it crosses the open spaces of roads and fields is distinctive and easy to recognize. Its song resembles that of the flicker, although the latter's cackle is shriller and faster.

The nest is a deep cavity hollowed out in a sufficiently large dead tree or

snag. The vertically oval hole is 3–4 inches in diameter, making it suitable for future use by a screech-owl. The eggs, like those of most cavity nesters, are white and round, but you won't find the shells at the base of the tree; that would alert predators to the presence of the nest.

The pileated woodpecker is fairly common in our towns, and relatively unwary. Feeding with little regard for human intruders, it often allows a close approach. Whether you view it with naked eyes or with binoculars, the sight is always spectacular.

Both sexes have solid black wings, tail, and body with a bright red crest. The male (right) has a red mustache behind and below the bill, while the female (left) has a black mustache.

Red-shouldered Hawk *Buteo lineatus*

The red-shouldered hawk is a buteo, member of a group of soaring hawks with short wings and stout bodies. It has longish wings and a longish tail for a buteo. It flies somewhat like an accipiter, with flaps and glides, though it may also soar like a red-tailed hawk. As soon as you hear its *kee-yar, kee-yar* or *kyear, kyear, kyear*, you know what it is. Unlike the raspy whistle of its cousin the red-tail, you can hear the red-shoulder's call for miles. Blue jays frequently imitate it, but the hawk's cry is louder and stronger than the blue jay's—and harsher and more fearsome if you are close to it.

Perched, the mature bird shows a reddish belly or a back with red-brown shoulder patches. In flight, the black tail barred with white is a dead giveaway.

Whereas the red-tail can be found just about anywhere, the red-shouldered prefers areas near streams, swamps, and moist woods. Its diet is mainly aquatic: frogs, turtles, and snakes, varied at times with rodents, rabbits, robins, screech-owls, crows, wasps, and grasshoppers. Though colloquially termed "hen hawk," it seldom takes poultry.

The mature bird shows a reddish belly and narrow white tail stripes.

Red-shouldered hawks stay paired for many years, possibly for life; the record is 26 years. The nest is built of twigs, bark, leaves, and softer things, and generally resembles that of the red-tailed hawk but is smaller. Pairs may return to the same nesting site season after season.

The reddish shoulders and horizontal barring on the breast are reliable field marks (opposite).

American Crow *Corvus brachyrhynchos*

Fish Crow *Corvus ossifragus*

When crows wheel and circle overhead, you can't help but admire their flight. And they're smarter than many birds: they can count as high as three or four. Both species are notoriously wary, and both are common on the coast.

Crows are black from head to feet with a big, thick bill. They vary in size. Although the American crow is slightly larger, the two species are difficult to tell apart visually. In order to distinguish these look-alikes, you must go by sound. The common crow says *caw, caw* or *awk awk awk awk*. The fish crow has a very nasal *car-ar*. Alas, the summer juveniles of common crows can sound a lot like that! But if your bird says *caw*, then for sure it's an American crow.

Both species are two or three times as large as any grackle or blackbird you'll see, but much smaller than the vultures that soar in the sky. Novice bird-watchers sometimes confuse a crow with a hawk. Crows are solid black; their beak is straight, not hooked; and they tend to sit on the very tops of trees instead of a little way down. Crows sometimes gather in flocks of hundreds, but it's more common to see them in groups of four or five.

Crows are omnivores with many ways of making a living. They'll eat grain, carrion, or anything else that opportunity presents (peanuts in your backyard, eggs in a nest, etc.). This bodes well for these adaptable species.

Both crow species build their bulky nests high in trees, often in pines. Great horned owls and red-tailed hawks sometimes use abandoned crows' nests.

Crows do birders a favor by letting them know when owls and hawks are nearby. They gather in numbers (5–25) to "mob" these predators, carrying on for 10 or 15 minutes, cawing harshly and stridently, and dive-bombing the hapless raptor. Keep your ears cocked for mobbing crows, and you'll find owls and hawks.

Crows are two or three times larger than any of our grackles or blackbirds.

No other solid black bird in our area is as large as
the crow and has eyes as dark as its body.

Cattle Egret *Bubulcus ibis*

Around 1880, the cattle egret made its way from Africa to the northeastern part of South America. How it covered that enormous distance is not clear! Sixty years later it reached the North American continent and began rapidly spreading through all the states and into Canada. It is now a common bird in the southeastern coastal states and California, and is found sporadically in most other states, where it feeds along roads and highways and in fields and pastures.

You can identify a cattle egret by its solid white plumage and yellow to orange bill and legs. The head, breast, and back take on a buffy or yellowish-orange color during the spring and summer breeding season, and the legs become pink.

The cattle egret is distinguished from the snowy egret by its bill and leg color (the snowy has a black bill and black legs with yellow feet) and stockier body (the snowy has a delicate, slender appearance). It is only about half the size of the great egret, which is also white with a yellow bill. In addition, the cattle egret usually feeds in dry areas rather than wetlands, typically on agricultural land, where it eats insects that are kicked up by livestock or tractors. It will eat frogs, snakes, crustaceans, and other tasty tidbits when given the opportunity.

The cattle egret is migratory and moves southward during the winter.

The arrival of cattle egrets in the States has had little impact on our other native species because the newcomers feed (and sometimes also nest) in dry lands rather than the wetlands where the other herons and egrets reside. Cattle egrets occasionally use rookeries that other herons or egrets might use.

Note the orange-brown chest of the breeding
plumage (opposite). The nonbreeding bird
(above) is pure white and has an orange bill,
similar to a great egret, but it is smaller in size
and has a shorter bill.

Red-tailed Hawk *Buteo jamaicensis*

The red-tail is common and widespread over the entire United States, soaring and hunting widely in open country. This buteo is quite successful as a bird of prey, as evidenced by its healthy numbers and many subspecies. As long as humans don't poison the rodents it eats, this hawk will get along just fine.

You may see the red-tail perched along the interstate, white breast shining in the sun. Or you may see it gliding into the grass of the median for a kill. It is dark brown above with white underparts and usually a dark brown breast band. Adults of both sexes have the reddish tail. The immature, however, has a drab tail banded dark and light with gray or brown.

The red-tail finds a tall tree in which to build a flat nest of twigs and sticks that is more than 2 feet in diameter. The pair continues to bring new green sprigs as long as the eggs are incubating. During this time the male also brings food to his mate. Nests may be reused.

It takes some practice to tell a soaring red-tail from a vulture. They're about the same size and soar in the same way, sometimes even together. But a vulture is black, and a red-tail will appear brown. The turkey vulture holds its wings in a dihedral—a shallow U or V—shape; the hawk holds its wings flat. If you can view the undertail against the sunlight, you'll see the hawk's reddish tail feathers showing through with a faint pink: definite identification.

The red-tail may hunt from perches near open fields or from the air. It may perch high—two-thirds of the way up a tree—or as low as a fence post. It wants small rodents such as rats, mice, rabbits, and squirrels but will take snakes and eat roadkill. On rare occasions it takes a bird. Chickens? No. This is not a chicken hawk. Help discourage farmers from shooting this beneficial bird: it eats the rats and squirrels that steal his corn!

Immature birds have stripes on their tail (barely visible here).

The red-tail's call is a rasping, one-note whistle that sounds almost like steam escaping.

The heavy body and short tail are typical buteo traits. The undertail faintly washed with red, dark hooded head, and dark breast band are reliable field marks. Our hawk's tail color is about that of a robin's breast and is easily seen against a blue sky (above).

Great Horned Owl *Bubo virginianus*

The great horned owl generally calls with five soft hoots: *hoo-huh-hoooo, hoo-hoo*. At night you can easily tell it from any other owl by this song alone. This is the largest owl in our area. It is named for its "horns," which are actually nothing more than tufts of feathers. Look for yellow eyes (the nonhorned barred owl has black eyes) and a prominent large, white bib under the throat. The light breast is barred with horizontal striping (the barred owl has vertical striping). The back is mottled brown and gray. There are 12 subspecies north of Mexico, each with its own slight color variations.

The great horned uses its wickedly sharp talons to capture anything that moves from dusk to dawn: squirrel, skunk, bird, rat, and snake. As is true of many birds of prey, the female is larger than the male. Her size gives her a little more control and protection during the mating process, and makes her a proficient hunter of larger prey.

The great horned roosts by day in deep shade, where it is often harder to see than you might expect, considering its size. The hiding place is sometimes disclosed by mobbing crows, whose frenzied chorus of caws alerts you to its presence.

Great horned owls begin breeding in late fall and lay eggs in early winter, often in an abandoned crow or hawk nest. The chicks, one or two, will hatch out during early spring. These birds get an early start because they need a long time to grow in order to survive the coming winter on their own. The fuzzy, gray chicks are surprisingly large. Their food-begging call, given about every half-minute, is a loud, rasping *screep*.

It is said that great horned owls probably exist in every county in the United States, even in metropolitan areas—that is how adaptable and widespread they are.

The "horns" may be relaxed and flopping to the sides or sticking up as if to make the bird look fiercer. The yellow eyes with black pupils, white bib, and horizontal barring on the breast are also field marks. The reddish brown facial disk is worth observing too.

Black Vulture *Coragyps atratus*

Turkey Vulture *Cathartes aura*

Turkey vultures and black vultures often soar and feed together during the day, though usually turkey vultures are more prevalent. Make sure to inspect any group of vultures to see if some of them look different from the others. The black vulture has an all-gray head, not the blood-red head of the turkey vulture. The black vulture also has whitish to light gray feathers only on the tips of its wings, while the turkey vulture has light gray from the wingtips to the body along the trailing edge. The latter characteristics are easily noted when the bird is soaring overhead. Farther away, notice that the black vulture flies with its wings straight out, while the turkey vulture's wings meet its body in a pronounced U or V. This V shape, which can be seen from miles away as the TV soars through the skies looking for a meal, also distinguishes it from the hawks—usually the red-tailed hawk—that soar in our southern skies. The tail of the black vulture is extremely short, so short that wings and tail seem to blend into one. Some people call the black vulture "short-tail."

Both species nest on the ground in some dark, secluded spot, and both feast on carrion. We see both species tugging at carcasses alongside interstates and other heavily traveled roads. Black vultures do not have a very good sense of smell and locate their food by sight alone or by following turkey vultures, which do have a good olfactory system.

Vultures rarely vocalize, so listening is not a useful way to distinguish the two species.

The turkey vulture has a red head (above top) and light gray
along the entire rear half of its underwings (above bottom).
The tail is wedge-shaped.

The black vulture (opposite left) has a gray head, and
only the tips of its underwings (opposite right) are light.
Note the short tail.

Common Yellowthroat *Geothlypis trichas*

"The black-masked yellow alarm system" might be a better name for this little bird. Come anywhere near his territory, and the male will come flying to a low bush and begin his low, harsh call note to let everyone around know you are there. He won't stop until you leave, either.

Every now and then the yellowthroat pops up and shows you his beautiful colors—yellow and black are what you notice first. His forehead and eyes have a wide black band across them down to the neck, almost like a Halloween mask. His throat and upper breast are bright yellow, and the black mask has a narrow, ash gray border above that blends into his brownish-green upperparts. His stomach is whitish and his legs are pink. The female lacks the black mask while retaining the other color pattern, including a white eye ring, but she is much duller overall so she can hide on her nest.

The common yellowthroat generally nests in open marshland. While the male is hopping about protecting the territory, the female is rarely observed until she begins feeding the young or during migration. The nest is low to the ground, typically almost on it or no higher than 3 feet.

You will see and hear these inquisitive little birds around marshes, ponds, streams, and other wetlands and habitats where there is low, dense shrubbery. You first know they are there by the male's loud singing—a very distinct *witchity witchity witchity*—coming from low bushes.

The yellowthroat's diet consists almost entirely of insects, which are gleaned from low bushes and leaves usually around the water.

As our wetlands are drained and suburban developments take more and more of their habitat, these charming little birds will continue to decline in numbers. They are also a common host for the ever-growing numbers of cowbirds that move in when forests are fragmented. Nevertheless, yellowthroats are still very common throughout the United States. They can be found year-round on the coast and in many wetlands farther inland.

The male displays his beautiful yellow and black pattern while protecting his territory. The much plainer female has a slightly yellow throat, olive-gray back and wings, and a white eye ring (opposite).

Savannah Sparrow *Passerculus sandwichensis*

Named for the city where a specimen was collected in 1811, this little brown bird is a very common resident of the Atlantic Coast, but only in the winter. In the spring it moves north to nest in the northern states and in most of Canada, coming back to us in late September or early October. As you do with any new bird, look first at its beak, which is a relatively short, stout one good for opening seeds, the major diet item for most sparrows.

Despite its common and widespread occurrence in Georgia, the Savannah sparrow is not nearly as well known as its similar cousin, the song sparrow (not included here). Both are small, brown birds with easily observed brown streaks on their white chests. The Savannah's streaks do not converge in a central spot as the song sparrow's do. The Savannah sparrow rarely sings its song on the southern coast. It is smaller than the song sparrow and stays more in field grasses, marsh edges, and meadows. The Savannah sparrow also has a

relatively short, notched tail; a pale yellow eye stripe; variable dark black stripes blended in with the brown of the back; and a white stripe down the center of its crown.

Look for small flocks of four to eight Savannah sparrows feeding in grasses along infrequently used roads, particularly those near water, but also in open fields and meadows. Males and females are not distinguishable.

We mostly hear its alarm *chip*, but this sparrow will occasionally sing its song—a weak, musical *tsit-tsit-tsit, tseee-taray*.

The Savannah sparrow has a short tail, streaked chest and flanks, and a stripe over the eye. The eye stripe can be clearly yellow (opposite top) or almost white (left).

Red-winged Blackbird *Agelaius phoeniceus*

The black males flaunt a red-orange epaulet at the bend of the wing that is a no-fail field mark. In the breeding season, this epaulet gets even redder and seems to glow, especially when the male spreads his wings in his mating display. Immature males seen in winter flocks have only a pale yellowish bar. Females are entirely different: all brown with a heavily streaked breast. You may mistake them for sparrows if you don't notice the large blackbird bill.

The red-winged blackbird nests in wetlands—beside streams and ponds, and in pastures and weedy fields. In winter, red-wings join other blackbirds to form vast flocks that roam the Southeast, foraging in fields and yards. Farmers consider them pests. Flocks can fill the sky like wisps of smoke. A large mixed flock of blackbirds foraging in your yard will feed by "rolling": birds at the rear fly to the front where there is fresh food; in this manner the flock inches forward.

The male (right) is all black except for the bright red shoulder patch, which can be broader than shown here and may include orange or yellow below the red. The female (left) has wide streaks on her underparts that are brown and tan rather than black.

In winter, red-wings join other blackbirds to form large flocks that roam the Southeast foraging in fields and yards.

How do you tell a red-wing in a big flock of black-colored birds? First, look for the red wing patch; then look at other details. Compared with the common grackle, the red-wing is smaller and has a smaller bill. In flight, it clearly lacks the grackle's long, wedge-shaped tail. The red-wing does not have the swept-wing jet look of a starling, and it is bigger than a cowbird and has a black head instead of a brown one.

In the breeding season, the aggressive and protective male produces a distinctive *oke-a-lee* song. In overhead flight, red-wings give a quiet *chuck*. The smallish nest is lashed to wetland shrubs and tall grasses. The young can climb and swim even before they can fly. Pairs may produce several broods in a summer.

Killdeer *Charadrius vociferus*

The killdeer, a kind of plover, is one of the few shorebirds that we regularly see inland. The killdeer is easy to identify by its two black bands on a white breast. The back and head are dull brown, but the rump shows a bright rusty color in flight. The killdeer's song resembles its English name: *kill-dee*. You can recognize it far away by its piercing piping or whistled *te-te-de-dit, de-dit*.

Open fields and pastures are the killdeer's favorite feeding grounds, but you may also see it by river and lake banks, and on the beach. Killdeer generally occur singly, searching the grass for worms, crickets, and grasshoppers, although loose feeding flocks of 10–15 are sometimes reported on Christmas bird counts.

This bird nests in gravelly places, including our driveways and flat rooftops. It makes a "scrape" in which the eggs lie unprotected on bare ground, though they are well camouflaged. The fuzzy young chicks (with just one neck stripe) are precocial; that is, they are able to run about and feed as soon as they hatch.

The killdeer is perhaps most famous for its "broken wing" behavior. The adult lures four-legged predators away from her nest by dragging her wings as if injured and piping plaintively. Just when the cat or fox is sure of an easy meal, the mother bird flies away. Other birds perform this trick too, though none is as famous for it as the killdeer.

Killdeer can also be found in mudflats.

The killdeer is often active at evening, making it a "crepuscular" bird. Listen for it after supper on your next camping trip.

The short, stubby beak and blunt forehead are typical of plovers. In addition to the two black breast bands, the killdeer has a brown back and wings, white underparts, and relatively long legs. Killdeer can be found in fields with short grass as well as in rocky terrains

Lesser Yellowlegs *Tringa flavipes*

Greater Yellowlegs *Tringa melanoleuca*

So similar are these two shorebird species in shape, color, and habitat that it is almost irrelevant to most bird-watchers which one they are viewing (unless, of course, one or the other is needed for a life list). The long, bright yellow legs are the first feature you notice. Both species also have grayish, barred upperparts in the winter (when we are most likely to see them); a streaked neck; and a slender, long, dark bill. The lesser yellowlegs is about 9 inches tall; the greater yellowlegs is a few inches taller.

Yellowlegs can be found in shallow wetlands, including isolated ponds next to supermarkets and parking lots, and in fresh or saltwater marshes and mudflats. They rarely show up on the beach.

Both yellowlegs species eat insects, crustaceans, and mollusks. They rarely probe into the mud, but instead snare small fish and any other suitable food items on the water surface.

Both species migrate to the far north to breed in the spring, but because of their staggered departure and return times, they seem to be around almost all the time.

The greater yellowlegs is a very wary bird. Hunters gave it the name "tattler" because these birds would spot them at a distance and immediately raise an alarm that would disperse large shorebird flocks. The lesser yellowlegs seems to be much less wary of humans and is almost tame by comparison. If you really want to tell them apart, check out the bill length. The lesser's bill is about equal in length to the head width; the greater's bill is about one and a half head widths, making it proportionally longer.

The greater's vocalization is a harsh series of three notes; the lesser gives softer single or double notes.

Both species have long, bright yellow legs. The key noticeable difference is the length of the bill in comparison to the head, longer for the greater yellowleg (opposite), about equal for the lesser (above).

Pied-billed Grebe *Podilymbus podiceps*

You usually see the pied-billed grebe as a solitary individual, or perhaps two or three at most, swimming warily in lakes, marshes, estuaries, and even sewage treatment ponds. At first glance it appears to be a small duck with an odd bill. It is not a duck, however; it is closely related to the loons. The pied-bill is totally at home in the water and dives to escape its predators rather than flying. That is also how it feeds, using its lobed toe pads and legs set far back on the body to propel it underwater to capture fish, crustaceans, frogs, tadpoles, and other aquatic morsels. In fact, the pied-billed grebe is among the best avian swimmers in the United States.

This grebe's small size (smaller than even the green-winged teal, not included here) initially alerts you that it is not a duck; its short, thick bill identifies it conclusively as a pied-billed grebe. In summer, the whitish bill has a black band around the middle and the throat has a black patch. The body is basically dull grayish brown with lighter but similarly dull underparts, although the rump is distinctly white.

The pied-bill is the most common and widespread of our American grebes and can be seen year-round in the southern states. While it likes freshwater ponds and lakes, it will feed in salt bays in the winter.

The pied-billed grebe's dense, waterproof breast feathers were popular in the turn-of-the-century millinery trade, and grebes were widely hunted for them, but not for food. Their muscles are coarse and unpalatable to all but raptors and alligators. While their waterproof plumage keeps them from becoming wet, grebes can adjust their buoyancy with ease, allowing them to submerge until only the head or eyes and beak are visible above the surface.

Summer, breeding plumage (top) features a black ring around the bill and a black neck patch. The winter pattern (bottom) features a faint bill ring and no neck patch.

Bufflehead *Bucephala albeola*

Our smallest diving duck is seen along the coast only in the winter. The bufflehead likes cold water and prefers to stay just south of where the ice forms. It breeds mostly in Canada near freshwater ponds, lakes, and rivers. During winter, buffleheads may frequent saltwater estuaries. The common name derives from this duck's "buffalo head" appearance when it puffs out the feathers on its oversized head.

The male is black above and white below, and the large white area behind each eye contrasts conspicuously with the rest of the dark head. While generally it looks black, the head is iridescent purple and green under some lighting conditions. The female is much duller: drab gray-brown above and with only a small white patch behind the eye. Both sexes have a gray or bluish, typically ducklike bill that quickly distinguishes the bufflehead from the similarly colored hooded merganser, which has a thin, long bill with a "tooth" at the tip.

The bufflehead is a fat little duck, well insulated for diving in cold water. That characteristic has also earned it the name "butterball." It is far less tasty than a Butterball turkey, however, and is not hunted as much as other ducks.

The bufflehead feeds on insects, snails, small fish, shrimp, and mollusks, foraging frequently in very shallow water with only the head submerged. They generally are found in small groups or pairs, and when disturbed can leap straight up and fly rapidly away. Other diving ducks must run along the water surface to reach flying speed before taking off.

The male (bottom) has a distinctive black-and-white pattern on the head. The colors of the female (top) are less conspicuous, and she has only a small, but distinct, white patch behind the eye.

Common Moorhen *Gallinula chloropus*

If you see this bird while it is swimming, you might think it is a dark-colored duck. On the other hand, if you see it walking and feeding along some freshwater pond, you might wonder if it is a dark chicken that has escaped the coop. Europeans thought so, and gave it the name "moor hen." If you look more closely, the prominent red bill and forehead immediately identify the moorhen, although in winter the red of both parts becomes brown. You may also notice that the back is dark brown and is separated from the black underparts by a white line along the flanks. The legs are greenish yellow.

The moorhen belongs to the rail family and, like its relatives, lives in freshwater marshes, ponds, and lakes. Unlike the rails, it swims well, constantly bobbing its head fore and aft as it paddles along. It resembles two other species: the American coot and the purple gallinule (not included here). The moorhen's bright red bill distinguishes it from the American coot, which has a white bill; and its brown back and black body distinguish it from the purple gallinule, which has an iridescent blue-purple body and a greenish back. The purple gallinule's blue forehead and lack of white on the flanks help to identify it as well.

The common moorhen feeds on aquatic and terrestrial plants, berries, insects, snails, and tadpoles, and can dive under the water briefly to reach them. It is one of the least shy of the rail family and goes about its business making all kinds of harsh but interesting sounds, including clucking, rasping, and cackling noises. Its long toes help it to walk well on the surface vegetation of ponds, but it flies—ungracefully—when it must.

While common moorhens can be found throughout the eastern United States in summer, in winter they reside mostly on the Atlantic and Gulf coasts, one of the reasons why many older books call this species the Florida gallinule.

The bright orange-red bill and forehead easily separate
common moorhens from coots and purple gallinules.

Blue-winged Teal *Anas discors*

The Southeast is a great place for birds because of its varied landscape, which includes a long coastline, open fields, mountains, and woods. Southern birds breed here, northern birds winter here, and many birds pass through on their way to and from breeding and wintering grounds. Thus, most of the birds found in the eastern United States are here at one time or another. That includes the blue-winged teal, which spends its winters here after breeding in puddle ponds throughout the Midwest. These teal arrive so early in fall—late August or early September—and leave so late in spring—late March into April—that they seem to be here all year long.

Blue-wings are small ducks that fly fast. They like to swim and eat in small, shallow ponds both inland on fresh water and on the coast in brackish marshes. The male is mottled brown with a blue patch on each wing that is not always visible. What is distinct, and what clearly identifies him, is the large white crescent just in front of each eye, extending from the forehead to the chin. You

The female is difficult to distinguish from several other species such as the less common green-winged teal (not shown) but does show a nice horizontal stripe through the eye.

The white vertical stripe in front of the male's eye is very distinctive for identification.

can't miss it. The female lacks the white crescent but also has a mottled brown body and wings, although less colorful than the male's. She has the blue wing patch too, easy to see when she is in flight but not always visible when she is paddling around; generally she is seen with a male.

Blue-winged teal migrate and feed in small groups of two to six birds. They eat pond weeds, seeds, and grasses. Under protected conditions, they can become almost tame and can be seen up close, but more often you see them at a distance near bushes and tall weeds or grasses on the other side of open, shallow ponds and marshes.

Clapper Rail *Rallus longirostris*

This common, relatively large rail is much more often heard than seen. It has a grayish, streaked back flanked by dull rufous patches on the wings and breast; the belly is marked by bands of gray and white. The head is mostly gray, and the bill is yellowish. The clapper is usually seen in the early morning or late evening, and under those poor lighting conditions simply looks like a grayish chicken with long legs and a long, curved bill. Some people call it the "mud hen" or "marsh hen" for that reason.

The clapper resides permanently in coastal saltwater marshes among the dense grasses and reeds, where it moves quietly along the mud at the water's edge, gliding in and out of clumps of vegetation. It can also be found in some Florida mangrove swamps. When out of sight in the marsh, the clapper gives its loud, cackling call—the notes varying in speed—and several other clapper rails typically answer it. It is difficult to convey the call in words or letters, but once you have heard this harsh sound coming out of the marsh, you won't easily forget it.

If the tide is out and you can see a clear stretch of mudflat, then wait patiently and quietly, and you may be rewarded with a view of this bird probing the mud for crustaceans or picking small items from the grasses as it moves through them. It will also eat frogs, fish, and some plants.

The nest is located on the ground near the high-tide level. It is made from reeds and floating debris that blend in beautifully with the surrounding area. The chicks are precocial—able to leave the nest as soon as they hatch.

Rails in general are dull-colored, secretive birds that are reluctant to fly if they can walk away through the reeds. When they do fly, it is typically just a short distance at low altitude. The rail's slender body fits between the reeds and gives rise to the expression "thin as a rail."

With a gray head and buffy underparts, a gray-brown body and a short tail, the clapper rail moves along like a long-billed chicken feeding in mudflats and reeds.

INCHES

Belted Kingfisher *Megaceryle alcyon*

Our belted kingfisher ranges widely over the entire United States. Lakes, rivers, and streams—even small ones—are its home. The kingfisher is distinctive in both coloration and shape. It has a conspicuous white breast with blue-gray back, wings, and crest, and a breast band of the same color. The female has a second, lower breast band of rusty red. The beak and head are large in relation to the body, and the crest is noticeable even at a distance.

Kingfishers feed by diving headfirst into the water, capturing fish which they swallow whole. They usually hunt from a perch near the bank. Occasionally one will hover like an osprey, and sometimes they will perch on power lines near water. After returning to the perch, the kingfisher subdues the fish and eats it. Although you have to pity the poor fish, it is certainly entertaining to watch as the kingfisher bashes its catch on a limb for a minute or two before swallowing it.

Kingfishers are solitary except when paired for nesting. The nest is a deep hole dug into the side of a clay bank. The sexes take turns digging, sometimes as far into the bank as 15 feet. When one mate comes to relieve the other on the nest, it calls from a nearby perch, then waits for the mate to leave before entering the tunnel.

The call—a thin, dry rattle—is usually given as the bird flies over the water in its characteristic erratic manner. Kingfishers love to fly up and down streams. If the stream is covered with a tree canopy, they fly as if through a tunnel, rattling as they go.

Spotted from c-102 marsh (8/2014)

MARSH AND POND

134

The male (top) has a single breast band. The female (bottom) has a second band that is rusty. Different lighting conditions produce different hues of blue for the back, but the large, shaggy crest is unmistakable.

American Coot *Fulica americana*

The coot looks more like a short-tailed duck than a rail, although it has rail-like feet with lobed toes for swimming. It has short wings, which is why it usually needs a little running room (on water, of course) to take off. The dark, slate-colored body contrasts with the ivory white bill, which is short and thick with a bare forehead shield. The coot's body characteristics are similar to the common moorhen's, but the white bill and forehead are distinct and clearly visible even at a distance and are readily differentiated from the moorhen's red ones.

The coot is one of the more prevalent species seen on freshwater lakes and ponds during Georgia and Florida winters. Sometimes coots flock together by the thousands. Lake Seminole, for example, at the western Georgia-Florida border, is home to nearly half a million coots in the winter. On the other hand, you can see small groups on city ponds or golf courses, and they can become very friendly when free food is handed out.

Coots eat most foods — plant material, aquatic animals, and even other birds' eggs. They usually forage while swimming on the water surface or just below it with the tail showing, but they can also dive for food or walk on the shore to get it. When swimming the coot nods its head in synchrony with its paddling, much as a moorhen does.

Unlike many marsh and pond birds, coots are very vocal and seem always to be tooting or cackling as if talking to everything around them.

The distinct white bill readily differentiates the coot from the common moorhen. The body can look dark or light gray depending on the lighting conditions, and the red eye is not always apparent.

Hooded Merganser *Lophodytes cucullatus*

Perhaps only the wood duck is more spectacular than the hooded merganser. The male hoodie has a black head adorned by a white fan-shaped crest bordered in black that is fully extended when he is excited or breeding. The back is also black, and the sides have fine brown streaks just above the point where the duck's body meets the water. The white breast has two vertical black stripes. The female is mostly gray, lighter in front, with a brownish hood. The hoodie's bill is slender and has serrated edges used for catching and holding its primary food—fish. In fact, the mergansers are the only ducks that specialize in a fish diet, although they will include frogs, crustaceans, and insects as dessert.

Hoodies breed and feed in wooded ponds, swamps, and streams in secluded woodlands. They can be seen on the coast mostly in brackish marshes during the winter, and move inland during the summer. They are rare nesters in Georgia, usually nesting in boxes put out for wood ducks.

The young are precocial and can feed themselves as soon as they leave the nest, although the mother hen does escort them around for about two weeks before leaving them on their own.

If you see them in flight, you can recognize mergansers because their head, neck, and body line up horizontally, straight as an arrow, as they move rapidly across the sky.

The male (bottom) has a large, white fan-shaped crest and a narrow, long beak. The female (top) has a brownish crest and a rather drab gray-brown body.

Wood Duck *Aix sponsa*

The wood duck is generally less common than the mallard—or at least much more secretive—but a far lovelier sight once you find it. The male is breathtaking. He has a crest that falls down over the back of his neck like a ponytail. The top of his head is green, his eyes and the base of his bill are red, his cheeks are black with white stripes, his breast is brown with fine light speckles, his sides are light gold, and his back and tail are black. All that finery is hard to see when he is in flight, though. Instead, look at the shape of the head: the bill will be down-turned. The less colorful female is brown with a wide, distinct, white eye ring. Listen for her alarm call, given in flight: a piercing *whoo-eek*.

Wood ducks almost became extinct early in the 20th century because people wanted their feathers for finery or desired stuffed specimens for display. The ducks have made a good comeback, due in large part to a program of putting nest boxes for them in ponds and swamps. You've doubtless seen the boxes: wooden structures on poles, usually protected from raccoons by a metal cone. Otherwise, the birds nest in tree hollows or large woodpecker holes, in which the female lays 10–12 white eggs.

The female's distinct white eye ring is a good field mark.

Wood ducks may walk quietly through the forest floor around their swamp, eating acorns. In the water, they dabble on the surface, eating insects and duckweed. They're found in the wildest and most beautiful places: thick swamps with moss draping off the branches. Since they have the ability to take off straight up, without spatting across the water, they can feel safe even in tight places. You sometimes find them on slow-moving creeks and beaver ponds. Usually they see you before you see them, so be prepared for their surprising burst into the air and their eerie cry as they vanish through the treetops.

The beautifully colored male (right) has a high forehead, a crest that looks almost like a backward-facing bill, and much white striping.

Green Heron *Butorides virescens*

Many kinds of birds eat the fish found in our freshwater streams, lakes, and swamps. Kingfishers, ospreys, and bald eagles (not included here) take fish right out of the middle of such waters. Egrets and herons hunt along the shoreline, going in only as deep as their long legs will take them. One short-legged fish eater stakes out the shallowest places of all, and sometimes even hunts from a dry limb just above the water surface: the green heron.

The deceptive thing about the green heron is its neck. When it is perched or coiled for a strike at its prey, it seems to have no neck at all. That is an illusion, for as it strikes or rises in flight, the neck stretches to a truly heronlike length. The chest and shoulders, which constitute half the surface area of this heron, are deep, rusty brown. The wings are a dark bronze that, with generosity and imagination, is sometimes called green. In breeding season, the male's legs are bright golden orange; otherwise the heron's legs are pale greenish or yellowish.

Green herons nest alone or in small colonies, building their nests anywhere from just a few feet to 20 feet above the water surface or the ground. They have mating rituals and displays as other birds do, including raising of crests and

The large bill attached to the long neck forms an efficient weapon for hunting.

plumes, crouching, and bill snapping. We are seldom privileged to see the green heron's courtship behavior, however, and should be quite satisfied just to see one fishing.

The loud call of this heron is distinctive and easy to recognize: a harsh, sharp *skeow* or *kyowk*.

The legs of females and nonbreeding males are pale yellow, but the male's legs become orange during the breeding season.

In its usual shoreline habitat, the green heron's back feathers may look bluish rather than green.

16

Mallard *Anas platyrhynchos*

The mallard is one of our most common ducks. Certainly it is the one most people can name because it thrives both in natural areas and in city and farm ponds. Individuals may stay around because their wings have been clipped, but most stay because they know where free food is!

The drakes, or males, of all our duck species have distinctive plumage. The mallard drake is easily recognized by his large yellow bill and green head with a white neck ring. You may also spot his recurved tail, from which we get the term "ducktail" (as in the haircut). After the mating season, the drake loses his bright colors and assumes a duller "eclipse" plumage. Female ducks, or hens, are usually drab brown and difficult to distinguish from one another. The mallard hen is a brownish bird with an orange bill marked with black, a dark eye stripe, and a blue wing patch, or speculum, that is present in both sexes but not always visible.

Ducks can usefully be divided into several groups. The mallard is a dabbler, or puddle duck, meaning that it feeds on the bottom of a shallow pond or lake by tipping its rear end skyward. It may also walk and feed on land. The pochards, or diving ducks, have legs set far back to help them swim and feed underwater, though this makes walking on land awkward.

The mallard offers an example of how bird-watching and conservation go hand in hand. Most of our ducks require freshwater ponds and wetlands for their survival. Americans have converted much of our wetland acreage into land for agriculture and living space. Such a practice, if continued, could lower the populations of ducks and other birds, taking some to extinction. For this reason many birders become avid conservationists, supporting public policies that protect habitats for wildlife. Mallards have benefited greatly from their efforts.

The drake (top) has a yellow bill, green head, white neck stripe, and a curled tail with black feathers that are clearly visible. The hen (bottom) has an orange bill with a dark top and a dark eye stripe. Both sexes have a blue speculum, although the male's is hidden here.

17

Barred Owl *Strix varia*

The barred owl is a large bird, although slightly smaller than the great horned owl. It haunts swamps and bottoms, eating small animals typical of wetlands. In size and habitat it stands to the great horned owl as the red-shouldered hawk does to the red-tailed. From the front, observe the strong vertical dark-and-white barring (hence the name) on the breast. See the round head (no "horns") and the black (not yellow) eyes.

The most exciting thing about the barred owl is its song. Known familiarly as the "eight-hooter," this owl's call is often described as *who cooks for you, who cooks for you-all?* But this basic song has many variations—all of them rather loud, some of them downright bizarre. The owl may sing just the first half of its song, or it may add or subtract a note along the way. Often it gives a single long, descending, guttural *hoo-aww*. When it sings in duet with its mate, it barks and caws. Owl-prowlers sometimes ask: Was that an owl or a dog? By contrast, the great horned owl's song is soft, usually consisting of four or five notes that sound like *who, who, who, who*.

Barred owls frequently nest in cavities and will also use an old hawk's nest or a properly designed nest box. Breeding and nesting begin in late fall or early winter, enabling the young to have a long growing season.

The barred owl hunts actively at night, in early morning, and even in the middle of the day. It may allow you to approach fairly close, and can be lured into view by the proper calls (usually recorded).

The dark brown vertical barring on the breast gives this bird its name.

The dark eyes and lack of "horns" easily distinguish the barred owl from the great horned. The owl can turn its head to face backward.

Snowy Egret *Egretta thula*

This medium-sized egret with its totally snow white plumage may be our most beautiful wading bird. As if to highlight these stunning white feathers, the bill and legs are completely black, but the feet are bright yellow, as is the bare skin at the base of the bill. The yellow areas on the face and beak become orange or red during the breeding season. Immature snowies have yellow streaks running down the back of their black legs.

Although little blue herons during their first year are also solid white, they have greenish legs and a gray bill with a dark tip. The all-white great egret is also somewhat similar to the snowy, but its much greater size and yellow bill quickly distinguish it. The cattle egret's yellow bill and legs likewise distinguish it from the snowy.

Snowy egrets can sometimes be seen in wet pastures eating insects alongside cattle egrets, but their primary habitats are coastal wetlands: marshes, ponds, rivers, and even beaches. They are year-round residents of southern coasts, but can be seen inland at lakes and ponds during the summer, particularly in Florida and southern Georgia.

Snowies eat an assortment of fish, crustaceans, insects, and small reptiles obtained while standing, walking, running, or stirring up the mud with their yellow feet. They roost communally and generally breed in large colonies called rookeries.

Much like those of the great egret, the plumes snowies produce on their back during the breeding season were once highly prized for ladies' hats, perhaps even more so because of their better size for that purpose. Thus, the adults were killed by the thousands at the peak of breeding, leaving the young to die. Once on the brink of extinction, the snowy egret was saved by protective laws and the efforts of the Audubon Society. It is now almost as common and widespread as it was in the early 1800s.

The snowy egret's entirely white body is accented by a
black bill and legs, as well as distinctive yellow feet.

Black-crowned Night-Heron *Nycticorax nycticorax*

This stocky, medium-sized heron with a black crown and back and all-white underparts is easy to recognize. The wings are gray, and the eyes are red. The night-heron keeps its head and neck tucked in close to the body, making it look even shorter and stockier than it is. The very pointed, shiny black bill looks like a dangerous dagger. The legs are short and yellow, turning to red during the breeding season.

Typically, you'll see this bird roosting in a tree or bush near the water's edge. It moves little during the day, then awakes and begins foraging at dusk and continues into the night. It would be interesting to know what visual system adaptations allow this species to find its food after dark in the water.

As the sun sets, the night-heron leaves its roost with strong wing strokes and unexpected speed and heads directly toward its hunting area. During this time it can be heard giving voice to a loud and raucous *quawk*. Its food consists of fish, frogs, snakes, insects, and sometimes the young of other wading birds.

You generally see a few black-crowns together, and sometimes they can practically fill a tree inconspicuously!

Unlike most other herons, immature night-herons have an entirely different color pattern from the adults: the entire body is mostly brown with white spots or streaks.

Although many of our coastal black-crowns are permanent residents, the species can be found across many of the nonmountainous regions of the United States during the breeding season. Birds that breed in northern areas migrate south for the winter.

The black crown and back contrasting with the light underparts identify this
medium-sized heron while it typically perches on a limb over a pond.

Little Blue Heron *Egretta caerulea*

While its name implies that this medium-sized bird should be close kin to the great blue heron, the little blue is actually more closely related to the snowy egret. In fact, immature little blue herons look very much like immature snowy egrets: both are white and medium sized, but the little blue has a bluish bill with a black tip, while the snowy's bill is all black. The little blue's legs are dull green, whereas the immature snowy egret has black legs with a yellow stripe up the back. Because the little blue requires two years to achieve adult plumage, partially white or partially blue birds are not uncommon.

Adult little blue herons are slate blue all over and have greenish legs. The bill is two-toned—mostly blue, changing to black at the tip. The long neck is held in an S-shape much of the time, both when the bird is fishing and when it is flying. In breeding season, the head and neck are reddish purple. At this time the little blue can be mistaken for a reddish egret. The latter has a similar color pattern, but its neck is quite shaggy in appearance and its legs are blue-gray. Because of its darker plumage and lack of many plume feathers, the little blue heron was not hunted for the millinery trade.

The white, immature bird could be mistaken for a snowy egret except for the greenish legs and two-toned bill.

The adult has a beautiful blue body and slightly reddish or purple neck.

Generally, little blue herons can be found in small groups in marshes, swamps, ponds, and wet fields. They forage while walking slowly and methodically in shallow water, looking intently for fish, frogs, crustaceans, and small snakes. They will also catch insects in fields. They are year-round residents of the coast, but wander inland and northward after the breeding season until winter drives them back to the southern coastal regions. Little blues prefer fresh water but will feed in salt or brackish water.

Tricolored Heron *Egretta tricolor*

Always look for a white belly contrasting with blue upper parts, and you'll have no trouble identifying this heron and distinguishing it from other herons. A strip of white runs down the front of the long neck, the bill is yellowish to blue-gray with a black tip, and the long legs are yellow to gray. Juveniles have chestnut on their back and neck, giving them an even more tricolor appearance.

The tricolor, formerly known as the Louisiana heron, is common on and near the Atlantic and Gulf coasts throughout the year. Unless it is breeding season (April to May) and you are at a rookery, however, you will probably not see tricolors in groups. They are not very gregarious and prefer to feed alone on their own little territory. They wade in shallow water as deep as their breast, stalking mostly fish, with some frogs, insects, and crustaceans thrown in for dessert. Sometimes they pose rock-still before a lightning-quick strike at a passing fish; at other times they step rapidly through the water, poking at everything almost indiscriminately. You usually see them in tidal flats and ponds, although occasionally they journey inland to lakes and streams.

Tricolored herons are less afraid of humans than most other herons, and you can get relatively close to them. Perhaps this is because they were never hunted for their feathers and plumes as many other heron species were. Apparently, their colors were not quite right for ladies' hats. For the same reason, their numbers did not plummet precipitously at the turn of the century. They are still abundant, even expanding their range northward to some extent, although there have been local population declines. Tricolors do not wander north and inland after breeding as much as other herons do.

The tricolored heron is easily identified by the contrast of its white belly against its beautiful grayish blue body (opposite). Note the reddish feathers at the base of the neck of the bird below.

White Ibis *Eudocimus albus*

It's hard to miss that big, down-curved bill regardless of whether an ibis is flying, feeding, or stationary. Couple that orange or red bill with a nearly solid white body and wings, and you have identified the white ibis. In flight, ibises display their distinct black wing tips. They also have long red or orange legs—more red during the breeding season—and similarly colored facial skin. These medium-sized wading birds are very common in summer. Typically, there are immature birds hanging out with the adults, and the youngsters will have a brown head, back, and wings with mostly white underparts.

White ibises occur year-round on the Atlantic and Gulf coasts, but are far fewer in the winter, when many head farther south. They feed in shallow waters of fresh or salt marshes and even in fields, gobbling up crustaceans (crabs, crayfish, etc.) primarily, but also insects and worms. Unlike herons, they like to feed in tight groups. Their nesting habits are similar, however, and hundreds or thousands gather in rookeries located deep within swamps or wetlands.

Of course, as wetlands are drained or otherwise developed and lost, the ibises lose some of their feeding and nesting grounds. While they remain numerous, their numbers have declined over the last 50 years or so. The audacious fish crows that rob their eggs don't help the situation, but the ibises generally lay more eggs when that occurs.

Like many of the birds seen near the coast, white ibises do not vocalize much, so their calls cannot easily be used for identification. The beak is so distinct, though, that you don't need anything else to identify this bird.

The immature bird has blotchy brown upperparts and a white belly.

The long down-curved orange bill is the key identification part (top). During breeding season, the bill becomes black and red along with the legs (bottom).

Osprey *Pandion haliaetus*

The osprey is a true fisherman, living nearly entirely on fish that it catches by diving feet first from as much as 100 feet or even more and riveting them with its sharp talons. That behavior gives this bird its other name, "fish hawk." It can hover like a kestrel over the water, and with its keen eyesight, pick out fish near the surface to dive on. After a catch, it exits the water on powerful wings, adjusts the fish so that its head points forward, and then flies back to its nest or perch. Widespread use of DDT once greatly reduced the number of ospreys, but they have now recovered.

A broad, dark line that runs through the eye to the back of the white head is one of the best identifying features once you realize that this bird is not an eagle hunting over the water. Although the wings arch like a gull's in flight, they also form a wide, shallow M that is another noteworthy characteristic. The osprey has a dark, brownish back, but is white underneath. The wrist has a black patch that is easily visible and distinctive for this species if viewed while it is soaring overhead. The tail is banded. Immature ospreys have lighter brown wing tips and back and a streaked crown for the first two years.

As this bird perches with its captured fish, it exhibits its
distinct white-and-brown facial pattern.

Ospreys usually flap their wings more while hunting than other hawks such
as the red-tailed, which likes to catch an updraft and sail over its territory rather
than expending energy to flap its wings.

Wherever there is an abundance of fish, be it salt or fresh water, ospreys can
usually be found. Except on the Southeast coast, they migrate during the winter
to warmer areas.

An interesting feature not usually noted is that the osprey's outer toe is
reversible in its direction of use, much like an owl's foot. Also, the leg feathers
reach only to the knee, and the feet are horny or scaly to more easily hold onto
slippery fish.

The osprey's voice consists of high-pitched whistles or down-slurred chirps
given in alarm.

The flying bird displays the shallow M and the black
patch at the wrists.

Canada Goose *Branta canadensis*

There is no mistaking a Canada goose. No other goose has the long black neck with the stark white underchin. No other is so loved by bread throwers at city ponds and parks, or so hated by lakeside homeowners who must deal with their excrement. On the other hand, they're said to be quite tasty.

In the 1970s, Canada geese were no longer coming to the Southeast because they were stopping in the mid-Atlantic states to feed on the abundant crops being grown to attract waterfowl for hunting purposes. The state decided to bring in goslings to help restore the population, and this worked well. However, one unfortunate result of this method was that the geese no longer migrated. The population stayed, bred, and now has become a nuisance on many lakes. This is a small price to pay for repatriating the species here, particularly as we can control their reproduction using various methods (egg turning, relocation, etc.), whereas an extirpated species has no options.

Canada geese mate for life. They are wary, strong, and protective of their young. Do not come between the parents and their goslings! An enraged goose is a fearsome bird.

The distinctive *ka-runk, ha-lunk* or *honk-a-lonk* calls of migrating geese are familiar sounds in the fall as their v formations pass overhead going north to south.

Ornithologists divide the Canada goose into subspecies, each with its own distinctive characteristics. Size, breast color, and neck ring all bespeak variations in the genome passed from generation to generation. Wild populations differ greatly. How else would they ever have evolved into the multifarious forms we see around us?

Everywhere

The male and female look the same.

The black neck and head and white chin strap and cheeks are unmistakable.

Anhinga *Anhinga anhinga*

The scientific name of the anhinga is easy to remember: the genus, species, and common name are all the same. It belongs to the family Anhingidae, so that should be easy to remember too. The anhinga has two other common names in the Southeast: "water turkey" and "snakebird." The former, apparently, is derived from the anhinga's long, wide tail; the latter derives from its ability to swim mostly submerged with only its long neck and head out of the water, looking somewhat like a water snake.

Although it is similar in appearance to its relative the cormorant, the anhinga has a longer neck and tail than the cormorant, and a very pointed yellow bill unlike the cormorant's hook-tipped beak. The long, sharp bill is used to spear fish during underwater fishing trips. The legs and feet are orange-yellow; the toes are webbed for swimming but have sharp claws for climbing out of the water.

Females display their brownish necks. The sharp-pointed bill for both sexes helps distinguish this species from cormorants.

A male bird demonstrates the typical wing-drying pose after swimming for fish.

Partial or full submersion is easy for the anhinga because it lacks the water-repellant oils found on the feathers of most other water-adapted birds such as ducks. Without these oils to make it buoyant, the snakebird can more easily dive, swim underwater, and catch fish. It becomes waterlogged after about 20–30 minutes of swimming, however, and must then climb out of the water onto a branch or log to spread its wide wings and dry its feathers. When it does so, it displays a broad white or silver band of covert feathers across the back of its wings that stand out beautifully on the male's nearly solid black body. The female is similar, but her tan head, neck, and breast distinguish her from the male.

The anhinga flies well by flapping, then gliding, looking like a "cross" against the sky because of its extended neck and long tail, and it readily soars to great heights. It is not a marine bird, and unlike its occasional roosting mates—the herons, ibises, and cormorants—it occurs throughout the coastal plain and may be more common near the coast in freshwater swamps, lakes, and rivers.

Great Egret *Ardea alba*

In general, egrets and herons are the same type of wading bird, with egrets being white, medium-sized herons. The great egret fits this description exactly: its plumage is all white, and it is about 6 inches shorter than the great blue heron. The great egret can certainly be confused with the cattle egret, but look for dark legs; a long, pointed, yellow to orange bill; and a slender, tall bird rather than the shorter, stouter, yellow-legged cattle egret. The great egret is also considerably taller than the snowy egret, which is clearly distinguished by its black bill and yellow feet. Like all herons and egrets, the great egret holds its neck in an S-shape most of the time, both standing and flying.

In summer after breeding, great egrets disperse widely throughout much of the United States, but in winter they return to the coasts.

During breeding, great egrets develop long, beautiful plumes called "aigrettes" that were so highly sought after to adorn ladies' hats that they were worth more than gold at the turn of the 20th century. Plume hunters almost wiped out the species, which previously numbered in the tens of thousands. Having saved this species (and others) from extinction by actually posting armed security guards around their roosting sites, the National Audubon Society adopted the great egret as its symbol.

Great egrets are not overly shy birds. I once watched one walk right into a fishing supply shed with the owner inside and start eating the minnows in a holding tank; it was probably a very hungry juvenile.

When alarmed, the great egret can produce a rattling croak, but generally it goes quietly on its way, spearing fish in the shallows or insects in grassy fields.

"Egret Tree"
field, feeding,
visible in
march 2014
S. from Cottage
1 cm (21 SH+)

Both the nesting bird and the beach-walking bird show the solid white plumage of this species, along with the daggerlike bill and long, dark legs.

INCHES

Until the 1970s this bird, our only native stork, was known as the "wood ibis," but it is a true stork. Furthermore, it is more closely related to the vultures than to the ibises. The wood stork is a very large white bird, second in height only to the great blue heron among our waterbirds, with a dark head and heavy, down-curved bill. The dark head and neck are due to a lack of feathers, as is true for the vultures, but the wood stork does not feed on carrion. It moves about in shallow ponds, marshes, and swamps with a partially opened bill that snaps shut on contact with its prey. Its food is mostly fish but can include crabs, snakes, turtles, crayfish, and some plant material.

Wood storks sometimes soar to great heights on warm days. When the bird is in flight, the black trailing edge of the wing and the black wing tips and tail make a distinct contrast against the white body and front of the wings. The long neck is carried straight out, seemingly to balance the long, trailing legs.

Wood storks nest in cypress or mangrove swamps in gregarious flocks from South Carolina southward into Central America. After breeding, many disperse northward, but they return south to the Gulf states for the winter. A few winter on the Georgia coast as well. They can be heard to croak during the breeding season, but otherwise are generally quiet.

Severe declines in numbers since the 1990s have been due to habitat destruction and changes in water management, particularly along the rapidly developing coast of Florida. Because of this, breeding colonies and numbers are becoming more prevalent in Georgia and South Carolina, where there is still good habitat for storks.

MARSH AND POND

166

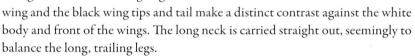

"Stork fea in marsh vued from 2010? rental on marsh. — full & bray. no longer severely endangered 2017.

This very large waterbird has a white body, large bill, and featherless head. The black along the wing edge (below) is much more prominent in flight; the feet are pink.

Great Blue Heron *Ardea herodias*

INCHES

The great blue heron is the largest waterbird on our coast. It has a 6-foot wingspan and stands 3 feet tall. The feathers are steely blue-gray, and the shoulders have a black patch. This heron is often seen standing at the edge of pond or stream with its long neck either held straight or curved into a graceful s. It flies with deep, slow strokes, legs extended and neck folded into a tight curve. Cranes and geese fly with their necks straight out.

With its neck extended, this bird is more than 3 feet tall, giving it an excellent view of fish. That size makes the great blue unmistakable even from a distance! The great blue is usually seen at the water's edge.

The great blue could be confused with the yellow-crowned night-heron (not included here), but that bird is smaller, shorter, and less common. The night-heron has a short black bill, whereas the great blue heron has a long yellow one. If you are seeing it from a good distance away, you might possibly think it resembles a wood stork or sandhill crane (not included here), but neither of those birds would be seen on the ground in our area.

The great blue feeds by spearing fish with its bill, tossing them in the air, and catching them headfirst for swallowing. Given the opportunity, it stakes out backyard fishponds and enjoys a gourmet feast of koi or goldfish. When alarmed, it flies off with a harsh grunt; otherwise it is silent. Look for great blues along the banks of rivers, in ponds of all sizes, and along marsh edges. The species' range includes much of the United States.

The nest is built of sticks in a wetland setting. In the last century and the early part of this one, the feathers of herons and egrets were prized as decorations for ladies' hats. Most of the herons mentioned in this book suffered population declines as a result of excessive hunting for their plumage, which in weight was worth more than gold.

MARSH AND POND

168

Also frequents Atlanta area. (2000's)

The great blue curls its neck during flight and while
preparing to spear fish with its daggerlike bill.

Least Sandpiper *Calidris minutilla*

The least sandpiper is one of about half a dozen species of small, grayish-brown shorebirds known as "peeps." It takes some practice and a good eye to identify many of these species, particularly in winter when their plumage may lack colorful and contrasting patterns. The least sandpiper, the smallest of the peeps, is never colorful. In winter, it has a dull gray back and head. The summer plumage displays a brown back and wings that give the least sandpiper a darker overall appearance compared with other small shorebirds. The underparts are essentially white with some gray and brown streaks on the upper breast. The yellowish-green legs separate the least from the western and semipalmated sandpipers (not included here), which have dark legs, although the legs of a least sandpiper may be covered with dark mud when it is feeding in the mudflats that it prefers. The dark bill is slender and pointed, and about the same length as the head width.

All of the peeps feed on sandy or muddy flats and eat small mollusks, crustaceans, insects, and marine worms as they walk along probing in the sand or mud.

Least sandpipers breed mostly in northern Canada and then migrate down to the Atlantic, Gulf, and Pacific coasts for the winter months. Many of the juveniles of this species do not migrate north to breed but remain behind during their first summer, so least sandpipers can be seen all year long on our coast.

The least, semipalmated, and western sandpipers are not easily distinguished in flight and even have somewhat similar calls: a *chet* for the semipalmated, *cheep* for the western, and *breep* for the least. If you aren't sure which sandpiper it is, don't feel bad, it takes a lot of practice to tell one peep from another.

This very small shorebird with mostly brown upperparts and white underparts is distinguished from the semipalmated and western sandpipers (not included) by its yellowish legs. The bird in the top photo is changing into its summer, breeding plumage; the drab, grayish-brown plumage (bottom) is that of winter.

Tree Swallow *Tachycineta bicolor*

The tree swallow is best identified by its iridescent blue-green upperparts and completely white underparts. Immature birds are white underneath as well but have brownish upperparts. The tail is only slightly forked, unlike the barn swallow's deeply forked tail.

Like all swallows, the tree swallow is an agile flyer that feeds in flight on insects over water or fields. It spends most of its time in the air and lands only occasionally, gliding effortlessly in a circle and ending with a short climb using a few quick wing flaps. Unlike most swallows, however, the tree swallow also eats the berries of bay trees and wax myrtles. The addition of plants to the diet allows the tree swallow to winter in more northern areas where insects are not always available, especially in cold weather, when

numbers of birds can sometimes be observed hanging on the branches of plants while feeding.

The tree swallow is a winter resident of coastal Georgia and Florida but is not usually seen here during the breeding season. Large flocks can be seen moving south down the Atlantic coastline in the fall. In the early spring, they head back north to nest. Tree swallows have nested in bluebird nest boxes in many locations near water across the northern United States and in Georgia.

The tree swallow's vocalizations consist of chattering or liquid twitters that begin with three descending notes.

The male has a bright blue-green back (opposite bottom; see also p. 3). The female is typically more drab (opposite top), while the juvenile has more brown on the upperparts (above).

Sanderling *Calidris alba*

No beach is complete without the small shorebird that runs behind the receding edges of the waves, poking its black bill into the wet sand and then racing ahead of the next wave as it washes onshore. The sanderling is always the bird closest to the ocean's edge and is usually the most active as it forages for its favorite food—small sand crabs that burrow in the wet sand in the shifting intertidal zone.

The sanderling nests in the Arctic tundras of the far north—about as far north as any nesting bird, almost next to the remaining ice—so we usually see this bird in its winter plumage. It is about the lightest colored of the sandpipers. The underparts are white, the upperparts are pale gray, and the legs are black. The dark patch at the bend of the wing distinguishes the sanderling from other peeps. In summer plumage, the head, back, and flanks are tinged with rusty brown. Sometimes first-year juveniles, which have buff on their breast, stay in the south rather than migrating to the northern breeding grounds.

Usually you'll see a small flock of these plump, gray birds moving together to and fro with the waves. When people approach them, they simply fly a short way out over the water and land a little farther up or down the beach to resume their sewing-machine tactics. During these flights, look for a conspicuous broad line of white running down the wings.

During migration, sanderlings have a few important staging areas where they stop and feed to gain the energy they need to carry them to their destination. Human encroachment on these areas has destroyed both habitat and food, such as horseshoe crab eggs, and this species had declined in numbers by as much as 80 percent since the 1970s!

The shoulder patch is also apparent, along with mottled back feathers, in juveniles.

The adult bird in breeding plumage has rich brown, black, and white feathers on its back and chest (top). Winter plumage (bottom) features a pale back and a black shoulder patch.

Semipalmated Plover *Charadrius semipalmatus*

Of the eight plovers found in North America, the semipalmated is the most abundant along our coast during migration in early spring and early fall. Many spend the winter along the U.S. coast, but most continue on to South America. You can identify the semipalmated plover by the short, blunt plover bill, which is orange with a black tip while the bird is breeding and solid black during winter. In particular, look for a single black neck ring: if there is one continuous ring, then you are most likely looking at a semipalmated plover. A killdeer has two black rings, and a piping plover (much less common and not included here) usually has a discontinuous ring and much lighter upperparts. The semipalmated plover's back and wings are brown; the legs are orange in summer and yellow in winter. The bands on the head and neck are brown in winter and thus match the back and wings more closely. The wings are typical of a plover—long and pointed—and the tail is short and rounded.

During migration, semipalmated plovers can be seen by the hundreds, dispersed widely over the beach as they hunt in the typical plover manner: sprint ahead, bob head while looking for food, sprint again. They feed on insects during the summer while they are breeding in the Artic and on small crustaceans, mollusks, and worms while they are wintering or migrating through here.

Although frequently observed on the beach, the semipalmated can also be seen in open areas of mudflats, marshes, and lakeshores. The call, *chur-wee*, is not often given in winter.

Juveniles have a brownish gray coloration on the upperparts.

A single complete neck band distinguishes this species. Breeding, summer plumage (top) features a more distinct band than the winter, nonbreeding plumage (bottom).

Ruddy Turnstone *Arenaria interpres*

The ruddy turnstone is among the easiest of the sandpipers and plovers to identify. The black or gray bib on each side of the throat and the bright orange or red legs are distinctive. This relatively small shorebird has a short, black bill like a plover's, but the bill comes to an upturned point as opposed to the usual blunt or fattened tip of most plovers. We generally see this short, chunky bird in its winter plumage as it roams the beaches, but we may see its rich reddish-brown back and wings and black-and-white head before it goes north to nest on the Arctic tundra. However, many juveniles remain on our coast during their first summer. In any case, the semicircular bib on the throat is distinctive. A dark stripe passes through the eye and another around the back of the neck. The sexes are similar, but the female's colors tend to be duller.

As its name implies, the ruddy turnstone runs around turning over or probing under stones, shells, and debris on beaches, mudflats, or rocky coasts searching for food. Its favorites are insects, mollusks, and crustaceans, but it will not turn down a meal of carrion, garbage, or even the eggs of other birds.

The turnstone is generally seen either in small groups or as isolated individuals. The species is found on the coasts of six continents in the winter. Because it nests on the tundra, well away from humans, it is not yet threatened by our ever-increasing population expansion.

The turnstone's voice is a sharp *tuk-a-tuk*, but it doesn't generally vocalize enough for birders to use that as a regular ID feature.

A richly colored male in summer plumage (above) features a clear black chest and head pattern and a plover bill. The winter plumage (left) features orange legs, a drab back, and a less-distinct chest and facial pattern.

Dunlin *Calidris alpina*

The dunlin is probably the most abundant shorebird on the coast during winter. Don't let the odd name (which means "little dune-colored bird") confuse you; this medium-sized bird is one of the many sandpipers. Its previous common name, "red-backed sandpiper," gives you a clue that the back can be reddish brown. The dunlin also has a very conspicuous black patch on the belly during the spring and summer, but the reddish color and spot are more often seen on the far northern tundras where the dunlin breeds. These characters can sometimes be observed on birds on our coast and even inland in early spring just before or during migration. The dunlin in winter plumage is an undistinguished shorebird with gray-brown upperparts and breast, white belly, and dark legs. The head has a light appearance with a very slight white eye stripe and chin. The most distinctive feature, if you are close enough to see it, is the downward droop of the relatively long black bill; it begins about two-thirds of the way along the bill. The bill is longer than a plover's but shorter than a willet's.

Like most sandpipers, the dunlin feeds on crustaceans, mollusks, marine worms, and insects. It likes to probe for these items on mudflats but enjoys the beach as well when food is abundant there. During fall migration, which takes place in late fall to winter on the Atlantic Coast, it can be seen on inland lakeshores or sewage ponds.

During the winter months, large flocks feed and rest together. When disturbed, dunlins perform a coordinated flight pattern that appears to have been carefully choreographed.

The dunlin has a distinct "drooping" bill. The black belly patch (top) makes clear the change toward breeding plumage, while its absence (bottom) defines a winter bird.

Least Tern *Sternula antillarum*

The least is the smallest of our terns, and size is thus one of the best features to use in identifying this bird. It is about one-half the size of a royal tern, and the body appears even smaller in comparison. The least tern has typically ternlike white underparts and a black cap with a white forehead. The back and the long, pointed wings are gray; the bill is yellow-orange with a small black tip. The tail is forked, and a distinctive leading black wing edge can be observed when the bird is in flight. This little tern hovers more than other terns as it spots its fish prey near the water surface and dives for it like a pelican. Like most terns, it keeps the head and bill generally pointed downward as it searches for food while flying lightly and gracefully over the water.

The least tern is seen in the United States only in the summer, during its breeding season; it winters on the coasts of Central and South America. It stays near water all the time, though, nesting on beaches and sandbars along the coast and along major rivers such as the Mississippi. Storm tides and water released from dams can wash away the nests of entire colonies because they are built so close to the water.

Least terns were hunted unmercifully during the late 1800s and early 1900s to adorn women's hats, and nearly disappeared before laws prohibited this incredible slaughter. Nevertheless, some populations are still endangered, and the others are monitored closely by conservation agencies. Fortunately, their numbers are increasing slowly as a result of careful protection and because they have adapted somewhat to human habitats by nesting on gravel rooftops.

Summer, breeding plumage (below) features a crisp black-and-white head pattern and a black-tipped yellow-orange bill, while winter plumage (opposite) features a dark bill and a mottled head pattern.

Black-bellied Plover *Pluvialis squatarola*

This plover, like most of its kin, has a rounded head and a short, straight bill that looks heavy and stout because it remains thick almost to its end. The black-bellied is the largest of our North American plovers, and you can pick it out relatively easily on the beach, mudflat, or open marsh, even at a distance, by its size and bill. Because it is also one of the shyest shorebirds, a distant view may be the only one you'll have.

In spring, before migration, the entire front of the bird, from its face to its belly, is solid black with a border of white that includes all the rest of the head; the upper body is spangled black and white. This "tuxedo" along with the erect posture and dignified demeanor make this plover the aristocrat of the beach. In winter, the posture and demeanor remain, but the underparts become mottled gray or white. In flight, which is fast and direct, this plover displays a white wing stripe, white rump, and black "wing pits."

Winter plumage features mottled gray-and-white underparts.
Note the short plover bill and white eye stripe. This species is
larger than most of the other plovers and sandpipers.

The black-bellied plover displays summer plumage as it
prepares for migration north, where it nests.

Black-bellied plovers may congregate in flocks to roost or migrate, but we
generally see them singly or in the company of one or two others. Because this
bird breeds on the Arctic tundra, we see it here mostly in winter, although
immature birds will spend their first summer in the South without migrating.
These juveniles can be recognized by their darker wings and back coupled with
pale yellowish spots and a streaked underbody.

The black-bellied plover does not usually probe in the sand or mud for its
food. It rapidly runs a short distance, stops dead, plucks at any surface food
it sees—the usual insects, mollusks, crustaceans, and occasional plant material—and then runs another short distance to repeat this feeding behavior.

Its voice is a somewhat mellow and musical *pe-oo-ee*.

Laughing Gull *Larus atricilla*

"Sea gulls" are as common for southeastern coastal residents as pigeons are for city dwellers, and perhaps even more so. In fact, there is no species or group officially known as "sea gulls," but there are many species of gulls. When we do apply the name, it should be only to species found on or near the sea. The most common of these is the laughing gull. This gull is relatively small compared with the less common herring gull and other rarer gulls such as the greater and lesser black-backed gulls (not included here). It is the only one that has a loud call that sounds like a strident laugh. You can hear these raucous birds just about anywhere near the coastline: on the beach and at piers, marinas, parks, fields, parking lots, and especially garbage dumps, where they forage for leftover human food.

Adult laughing gulls in their breeding plumage have a black-hooded head and

Adult breeding plumage (top) features a hooded black head, a charcoal gray back, and a red bill. Winter plumage (bottom) features a mottled gray head, a gray back, and a dark bill. Both sexes are alike.

black wing tips; the back and most of the wing tops are dark gray, contrasting with the white of the neck and underparts. There is also a narrow white eye ring. The bill is red or carmine, and the legs are dark. In winter, the head loses its hood and becomes whitish with a smudge of gray over the head. First-year juveniles are nearly all brown except for a white rump and trailing wing edge and a dusky breast.

Laughing gulls are very gregarious in their roosting, feeding, and nesting habits. Flocks of them can be attracted to bread or cracker handouts, and some will take food on the fly directly from your hand. Obviously, they have learned to live near humans and extract whatever food they can find from our hands, dumps, and parking lots. If you come bearing food, you instantly become a friend of this species. Once these gulls become your friends, the other species become much less wary of you, and thus may be much easier to see (and photograph).

A winter plumage adult stands while a juvenile, with its
molting brown back, rests on the sand.

Willet *Tringa semipalmata*

You see a large shorebird (about the size of a crow) walking nonchalantly along the beach, often well away from the water or in a tidal pool, picking up little tidbits of food. If it's winter, this bird will be plain gray above and whitish below, its only distinctive characteristic being its long legs, which appear to be blue-gray. If it's summer, the body will be heavily barred and the head and neck streaked with brown. The bird has a fairly long, straight bill that it uses occasionally to probe the sand or mud. Only when it takes flight will the bird display its best identification characteristic: as it raises its wings high over its head or flies off, its boldly colored white-and-black wings suddenly become prominent. The black along the trailing edge of the wings and the broad white stripe in front of it cannot go unnoticed: it's a willet!

Willets eat crustaceans—they're fond of fiddler crabs—and some plant material. The call is *pee-wee-wee* or the loud *pill-will-willit* that gives the willet its name. During the breeding season, willets can be downright noisy and long-winded, letting everyone know that this is their territory.

There are two subspecies of this bird: one breeds on the Atlantic and Gulf coasts; the other breeds on the western prairies around freshwater wetlands. While most birds of the prairie group move to the West Coast and south for winter, some meander over to the East Coast and become the major subspecies

here in winter because most of the Atlantic group birds migrate south.

In the early part of the century, both subspecies were hunted almost to extinction. Their numbers recovered after new laws stopped the relentless shooting.

Summer plumage (opposite) features a brown barred head and flanks. A winter plumage bird (above) lacks the brown barring, but it does feature the distinctive gray legs. Look for a solitary large bird feeding along the shoreline.

Forster's Tern *Sterna forsteri*

The most conspicuous feature of the Forster's in flight is the long, deeply forked tail that has earned it the nickname "sea swallow." When the bird is at rest, the long tail extends well beyond the wings. A second good field mark is a black patch behind the eye that is easy to see when the bird is station-

This bird is transitioning to its breeding plumage. Note the partially white forehead and the long wings that extend beyond the tail.

ary. This patch is present only in winter plumage, but that is the time of the year when Forster's terns visit Georgia. They breed mostly in northern marshes and along the western Gulf Coast. During the breeding season, the plumage features a black cap, gray upperparts, and white underparts; the bill is orange with a black tip. The winter bill is black, and the legs are orange-red in both summer and winter.

The Forster's is very similar to the common tern, but only the Forster's winters on the U.S. coasts. The common tern passes through during migration (April through May, August through October) and has darker wingtips. In total length the Forster's seems not very different from the royal tern, but when you see the two together, the Forster's tern is clearly much smaller overall than the royal (see photo on opposite page).

The Forster's tern plunges from the sky to catch surface fish, but it also picks insects out of the air or off the water surface.

Its call—*keer* or *kyaar*—might help you to identify this bird at a greater distance.

Although once hunted heavily for its long plumes, the Forster's tern has recovered well and is expanding its range northward along the Atlantic Coast.

The gray-white crown and the black patch behind the eye typically mark winter birds along the Atlantic Coast. Note the comparative size of Forster's tern to that of the royal tern behind it.

Ring-billed Gull *Larus delawarensis*

The ring-billed is one of the most common gulls on our Atlantic coastline during winter. In size it is between a laughing gull and a herring gull, but the distinct black ring around the yellow bill is the key feature to look for in identifying this bird. The underparts and head are pure white, and the back and black-tipped wings are bluish gray. In winter, the back of the head and neck will show some dusky spots. Unfortunately, first-year and second-year birds are very similar to immature herring gulls, including the pinkish legs. Both species are mottled brown and blue-gray, with similarly colored spots on the underparts. If you really want to identify the juveniles, look at the tail during flight: the ring-billed has a distinct black band, and the herring has a diffused broad band. Otherwise, it's an immature gull, species unknown.

Another clue, even from a distance, is location. If you see a gull (or flock of them) hanging around a fast-food restaurant, then you can bet that it's a ringed-billed. This species has adapted well to humans and their food, and can be found where scraps are easily obtained, such as garbage dumps, fishing boats, piers, and marinas, and, of course, McDonald's. Being omnivorous, they will also eat worms and insects from plowed fields, rodents, seeds, and other odds and ends that any opportunistic scavenger loves. On the bad side, they rob other birds, including taking their eggs and young, and often have a significant negative impact on populations of birds such as terns.

The adults spend their summers nesting mostly in Canada and the very northern edge of the United States. Many juveniles stay behind, however, and can be seen along our coasts in summer.

This gull's vocalizations are loud, raucous, sometimes high-pitched mewing sounds.

Juveniles are mottled brown and blue-gray, with similarly colored spots on the underparts. During flight the ring-billed juvenile has a distinct black band on its tail.

The winter ring-billed gull still carries a black ring on the bill but is much less colorful than during the breeding season. The sexes are alike.

During the breeding season, the ring-billed gull has very bright yellow legs, the distinctive black ring on the bill, and eyes ringed by red.

Boat-tailed Grackle *Quiscalus major*

You might easily mistake a boat-tailed for a common grackle—at least until you note its unusual tail, which is very long and v- or keel-shaped (thus the name). Like most blackbirds, the male boat-tailed is essentially all black; but like the common grackle male, his feathers are iridescent. When the feathers reflect light, the wavelengths of certain colors are exaggerated or augmented—in this case, to give a bluish or greenish sheen to the overall black color. This is true only for the male. Also, when you compare sizes, the male boat-tailed is much larger than the common. The female boat-tailed is considerably smaller than the male (closer to a common grackle in size) but also has the long tail. Her head, neck, and breast are brown, and her back and wings are darker and are not iridescent.

Boat-tailed grackles reside almost exclusively on the Atlantic and Gulf coasts and are rarely observed inland except in Florida, where they may live and nest close to inland marshes or lakes. They like to associate with others of their species, and so form gregarious colonies. Even when nesting, they usually can be found close together. Oddly, the western Gulf Coast subspecies usually has dark eyes while the Atlantic subspecies has yellow eyes.

Boat-tailed grackles are omnivorous. Their diet includes land insects, lots of crustaceans, and eggs and young of other birds. They will also wade into shallow water to eat small fish when they can't obtain easy handouts or leftovers from humans.

The voice is a harsh assortment of cackles and whistles along with series of *check*s.

The male (bottom) is solid black with a long V-shaped tail and a yellow eye. The female (top) also has a long V-shaped tail but is a mousy gray-brown and has smudged black wings.

American Oystercatcher *Haematopus palliatus*

This fairly large shorebird immediately catches your eye with its long, bright orange-red bill and black-and-white body. You may be confident of your identification, but double-check that you haven't confused this bird with a black skimmer, which has similar color features. The skimmer, however, has a lower mandible that sticks prominently beyond the upper mandible and is black at the tip, and the legs are red and very short. The oystercatcher's legs are long and pinkish, and its entire head, neck, and upper breast are black except for a red eye ring. These differences are clear, but the other characters are close enough to make you look twice before positively identifying each species.

Our oystercatcher is found strictly on the coast where it can find its favor-

ite shellfish: mussels, clams, and oysters. It generally spots its food visually, but can probe in the sand or mud and find mollusks and other live goodies by touch. The strong bill either quickly slips into a partially opened shell to reach and cut the adductor muscle or simply hammers a hole in the shell to do the same.

The American oystercatcher likes isolated little islands, sandbars, beaches, or mudflats on which to rest or feed. While it can stand qui-

etly or walk slowly and peacefully down the beach, it can also be noisy when a flock gathers, giving out loud *wheep wheep* calls.

Despite habitat loss along coastal areas, the oystercatcher has gained some human-made breeding grounds, as it will now use dredge spoil islands. In addition, the species is returning to the more northerly areas of New England where its range reached during the 1800s.

The oystercatcher is an unmistakable large shorebird with a completely orange bill, white underparts, and black and brown upperparts.

17

Black Skimmer *Rynchops niger*

While the skimmer could be mistaken initially for an oystercatcher because it is all black above and all white below, you can always distinguish the skimmer by its bill and legs. The skimmer's lower mandible is about an inch longer than the upper one, and its legs are much shorter than the oystercatcher's legs. The wing tips extend well beyond the tail when the bird is at rest. The skimmer's legs and bill are carmine red, and the bill has a black tip. Besides the irregular shape, the bill is also knife-edge thin. The skimmer flies with the lower mandible just under the water surface. When the lower bill contacts a fish, the upper one snaps shut, trapping the food. This requires an extremely fast reflex reaction, and it occurs as the bird skims perfectly smoothly above the water surface at about 5 mph on wings that can be as much as 4 feet from tip to tip!

Because it needs a smooth water surface to feed, the skimmer prefers sheltered bays and tidal inlets, lagoons, and estuaries, generally of salt water. It lives and breeds on the Atlantic Coast, mostly the southern part, and the Gulf Coast, and is rarely seen inland. Individuals occasionally nest on gravel rooftops, but skimmers are typically very gregarious, with hundreds nesting together in the sand of secluded beaches or offshore islands.

Another interesting feature of the skimmer is the pupil of its eye, which is vertical, like a cat's, rather than round like that of most birds, including the other waterbirds. This might be an adaptation related to its frequent feeding at dusk and night when the water is calmer and fish may be nearer the surface.

The skimmer is similar in size to the oystercatcher (previous page), but the bill is mostly black, the lower mandible extends beyond the upper mandible, the legs are shorter, and the long wings extend beyond the tail. The lower mandible is used to catch prey, slicing through the water until it encounters something, at which time it snaps shut, hopefully with a meal (top).

18

Royal Tern *Sterna maxima*

There is nothing particularly royal about this tern. In fact, it is one of the most common birds you'll see on the beach, along with the laughing and ring-billed gulls. It is nearly as large as those gulls, but its pointed wings and forked tail give it a decidedly ternlike appearance in flight, as does its hovering ability before it dives from the sky to take small fish. The bill is orange. The key characteristic visible when this bird is at rest on the beach is a spiky, black crest at the back of the head. In summer, the black crest extends forward to the bill and just below the eyes; in winter, the forehead is white. The grayish back and wings with white underparts don't help to distinguish this bird, as many birds along the coast have these characteristics, but the dark trailing edges of the wing tips and the dark outermost primary feathers help a little when the royal is observed in flight. The Caspian tern (not included here) is slightly larger, has a red bill, and generally is seen as one or two individuals rather than the crowd of royal terns that usually hang out together.

The royal tern occurs in Georgia year-round, but only on the southern Atlantic Coast. Inland appearances are very rare, probably only after strong winds have blown it there.

The royal's voice is a *keer* given fairly frequently in flight, so it is actually helpful in identifying this bird.

Although seriously threatened early in the century because birds were killed for the millinery trade and eggs were taken for food, the royal tern's numbers have rebounded very well.

Summer plumage features a solid black crown.
The spiky feathers on the back of the head appear
in all seasons, but the bill varies in color.

Winter plumage features a white forehead.

Herring Gull *Larus argentatus*

On southeastern coasts, the herring gull is the largest gull commonly seen during the winter (the black-backed gull being bigger but much less common, and not included here). The herring, in turn, is much less frequently seen than the numerous laughing and ring-billed gulls, although it used to be the most abundant gull on the Atlantic Coast before hunters decimated it early in the 20th century. Generally, you'll see one or a few individuals hanging around with other gull species, but rarely big flocks of herrings.

The herring gull's pinkish legs are distinct and identify it almost immediately. Attached to these are a pure white lower body, head, and neck; the back and wing tops have a bluish-gray mantle. The wing tips are black with white patches, and the bill is yellow and thick with a red spot on the lower mandible. In winter, there are brown streaks on the head and neck. The flat forehead and large bill give the herring a "mean" look that goes with its appetite and size. Immature birds are uniformly mottled brown; the bill is black with some pink. The legs are initially dark and change to pink during the first few months of life. The rest of the body changes to adult plumage over the next three years.

Herring gulls will eat anything edible. They forage for live or dead fish on the water surface and scavenge other carrion on the beach, leftovers at garbage dumps, baby birds, rodents, and any food that can be stolen from other birds.

They are also smart enough to carry shellfish high into the air and drop them on rocks to crack them open. While the herring is generally wary and naturally suspicious, it rapidly adapts to safe areas near humans, including piers, harbors, and ships.

The herring gull vocalizes during mating and nesting, when it is alarmed, and when it is communicating with other birds, but its most common call is *kyow-kyow*.

A juvenile (top) has brown plumage but still displays pink legs. As herring gulls change into summer plumage, the brown streaks on the head and neck (opposite page) begin to disappear (bottom).

Winter plumage for this gull (above) features a white body, gray wings, and a yellow bill with a red spot. The herring is much larger than other gulls, such as the ring-billed next to it (opposite).

Double-crested Cormorant *Phalacrocorax auritus*

This species was clearly named by someone who had a dead bird in hand and could see marks not usually visible on a live specimen in the field. You almost never see any crests on the head of this cormorant. It might be better named the "orange-faced cormorant."

Other species of cormorant are present in the United States, but the double-crested is the only one you are likely to see along the southern Atlantic Coast. It is a large, completely black bird with big, webbed feet. If perched, it stands upright. The bill is orange and has a curved hook at the end. If the bird is in flight, you will see a body much larger than a crow's with a long, extended neck and a long tail. It is sometimes called the "crow duck." You might mistake a cormorant for an anhinga, but the two species have different flight patterns. The anhinga generally flaps and glides, whereas the cormorant continually flaps. In addition, the anhinga's bill is sharply pointed. When a cormorant is in the water, most of the body is submerged with just the long neck and head visible. You might confuse it with a loon then (not included here), but cormorant's hooked beak is usually visible.

You commonly see cormorants perched on buoys, pilings of wharves or power lines, and trees close to the water. They pursue their prey (fish, crayfish, frogs, snakes, etc.) underwater, propelled strongly and rapidly by their webbed feet. Because, like the anhinga, the cormorant does not waterproof its feathers with oil from a special gland, the feathers trap less air. The lack of buoyancy allows for easier underwater swimming, but the feathers become waterlogged and must dry out afterward. That's why you often see cormorants perched with their wings spread.

The cormorant does not normally vocalize except when nesting, and then only makes grunts and croaks.

This solid black bird is similar to the anhinga but has a distinct hook at the end of the bill. The adult bird (right) is indistinguishable by sex.

Juveniles (top) have a brownish-gray neck and breast. While fishing, this species can submerge its body and show just the head and neck (bottom).

Brown Pelican *Pelecanus occidentalis*

After sea gulls, most people probably think of pelicans when they visualize the seashore. This large, odd-looking bird does not range beyond the coasts into inland freshwater sites, but stays close to its food source—saltwater fish. It feeds in fairly shallow water by diving from heights as much as 50–60 feet above the surface. Like the other six pelican species of the world, the brown pelican has a long, narrow bill with a large pouch stretching across the lower mandible. The pouch collects water and fish when the pelican strikes the surface. The water drains away, leaving a meal for the bird inside the pouch. Only the brown pelican dives from the sky to catch its prey; the other pelican species hunt while swimming along on the water surface.

As the name implies, the brown pelican has a brown body and dark legs that are simply dark when viewed in silhouette against the sky. The neck and head are white, and have chestnut on their back side during the breeding season. The head sits atop a long neck attached to a big, heavy body that manages to fly only because the large wings have a 6–7 foot wingspan. Immature birds have an all-brown head and neck.

Brown pelicans almost disappeared in the early 1970s because of the pesticide DDT. The fish the pelicans ate contained DDT, picked up from their surroundings, and the pesticide accumulated in the pelicans' bodies and caused their eggshells to become thin and brittle. They have recovered very nicely since the ban on most harmful pesticides. Probably the main threat to them now is fishing line. The pelicans take fish caught on fishing lines and become injured or entangled, making it difficult for them to survive (to a ripe old age of 25 years or more). They are the steadfast companions of fishermen, who usually end up sharing some of their catch (willingly or unwillingly).

While pelicans are clumsy and awkward on land, they are graceful and powerful over the ocean surface, where you often see them flying in line with their cohorts. Their low flight allows them to use the "ground effect" of lift to carry them along just above the water. Adults generally make no sounds and nest in large flocks on coastal islands.

Winter plumage for the adult features a white neck (top), while
a velvet brown neck is part of the breeding plumage (bottom).
Juveniles have a brownish head and body (opposite).

Checklist of Common Birds of Coastal Georgia

RELATIVE ABUNDANCE: C = common, seen in good numbers in appropriate habitat and season; FC = fairly common, seen in moderate numbers in appropriate habitat and season.

SEASONAL OCCURRENCE: PR = permanent resident, present year-round; SR = summer resident, breeds; WV = winter visitor, present in winter;

M = migrant, transient during spring/fall migration.

		ABUNDANCE	SEASON
○	American Coot	C	PR
○	American Crow	C	PR
○	American Goldfinch	C	WV
○	American Kestrel	C	WV
○	American Oystercatcher	FC	PR
○	American Robin	C	WV
○	Anhinga	C	PR
○	Barn Swallow	C	PR
○	Barred Owl	FC	PR
○	Belted Kingfisher	C	PR
○	Black-bellied Plover	C	WV
○	Black-crowned Night-Heron	C	PR
○	Black Skimmer	C	PR
○	Black Vulture	C	PR
○	Blue-gray Gnatcatcher	C	PR
○	Blue Jay	C	PR
○	Blue-winged Teal	C	WV
○	Boat-tailed Grackle	C	PR
○	Brown-headed Cowbird	C	PR
○	Brown Pelican	C	PR
○	Brown Thrasher	C	PR
○	Bufflehead	C	WV
○	Canada Goose	FC	PR
○	Carolina Chickadee	C	PR
○	Carolina Wren	C	PR
○	Cattle Egret	C	PR
○	Cedar Waxwing	C	WV
○	Chimney Swift	C	SR

		ABUNDANCE	SEASON
○	Chipping Sparrow	C	WV
○	Clapper Rail	C	PR
○	Common Grackle	FC	PR
○	Common Moorhen	C	PR
○	Common Yellowthroat	C	PR
○	Double-crested Cormorant	FC	PR
○	Downy Woodpecker	C	PR
○	Dunlin	C	WV
○	Eastern Bluebird	C	PR
○	Eastern Kingbird	C	SR
○	Eastern Meadowlark	FC	PR
○	Eastern Phoebe	C	WV
○	Eastern Screech-Owl	C	PR
○	Eastern Towhee	C	WV
○	Eurasian Collared-Dove	C	PR
○	European Starling	C	PR
○	Fish Crow	C	PR
○	Forster's Tern	C	WV
○	Gray Catbird	FC	PR
○	Great Blue Heron	C	PR
○	Great Crested Flycatcher	FC	SR
○	Great Egret	C	PR
○	Great Horned Owl	FC	PR
○	Greater Yellowlegs	C	WV
○	Green Heron	C	SR
○	Herring Gull	C	PR

	ABUNDANCE	SEASON
Hooded Merganser	C	WV/PR
House Sparrow	C	PR
Indigo Bunting	C	SR
Killdeer	C	PR
Laughing Gull	C	PR
Least Sandpiper	C	PR
Least Tern	C	SR
Lesser Yellowlegs	C	WV
Little Blue Heron	C	PR
Mallard	C	PR
Mourning Dove	C	PR
Northern Bobwhite	C	PR
Northern Cardinal	C	PR
Northern Flicker	C	PR
Northern Mockingbird	C	PR
Osprey	C	PR
Painted Bunting	C	SR
Pied-billed Grebe	C	PR
Pileated Woodpecker	C	PR
Pine Warbler	C	PR
Purple Martin	C	SR
Red-bellied Woodpecker	C	PR
Red-eyed Vireo	C	SR
Red-headed Woodpecker	FC	PR
Red-shouldered Hawk	FC	PR
Red-tailed Hawk	C	PR

	ABUNDANCE	SEASON
Red-winged Blackbird	C	PR
Ring-billed Gull	C	PR
Rock Pigeon	C	PR
Royal Tern	C	PR
Ruby-crowned Kinglet	C	WV
Ruby-throated Hummingbird	C	SR
Ruddy Turnstone	C	PR
Sanderling	FC	PR
Savannah Sparrow	C	WV
Semipalmated Plover	C	WV
Snowy Egret	C	PR
Summer Tanager	C	SR
Tree Swallow	C	PR/M
Tricolored Heron	C	PR
Tufted Titmouse	C	PR
Turkey Vulture	C	PR
White-eyed Vireo	FC	PR
White Ibis	C	PR
White-throated Sparrow	C	WV
Willet	C	PR
Wood Duck	C	PR
Wood Stork	FC	SR
Yellow-bellied Sapsucker	FC	WV
Yellow-rumped Warbler	C	WV

Image Credits

I provided most of the images myself, but for those that I did not, I thank the following sources:

p. xv: Alison Cundiff
p. 6 (top and bottom): Todd Schneider
p. 14: Tony Campbell / Dreamstime.com
p. 15 (bottom): Howard Cheek / Dreamstime.com
p. 19: Giff Beaton
p. 80: John Pitcher / istockphoto.com
p. 81: Megan Lorenz / istockphoto.com
p. 107 (top): Jerry Amerson
p. 114: Giff Beaton
p. 119: Rinusbaak / Dreamstime.com
p. 131: Giff Beaton
p. 140: Stubblefieldphoto / Dreamstime.com

Index

jay, blue, 86–87

kestrel, American, 80–81
killdeer, 120–21
kingbird, eastern, 52–53
kingfisher, belted, 134–35
kinglet, ruby-crowned, 6–7

Larus argentatus, 204–5
Larus atricilla, 188–89
Larus delawarensis, 194–95
laughing gull, 188–89
least sandpiper, 172–73
least tern, 184–85
lesser yellowlegs, 122–23
little blue heron, 152–53
Lophodytes cucullatus, 138–39

mallard, 144–45
martin, purple, 56–57
meadowlark, eastern, 74–75
Megaceryle alcyon, 134–35
Megascops asio, 70–71
Melanerpes carolinus, 78–79
Melanerpes erythrocephalus, 62–63
merganser, hooded, 138–39
Mimus polyglottos, 82–83
mockingbird, northern, 82–83
Molothrus ater, 48–49
moorhen, common, 128–29
mourning dove, 88–89
Mycteria americana, 166–69
Myiarchus crinitus, 58–59

night-heron, black-crowned, 150–51
northern bobwhite, 72–73
northern cardinal, 66–67
northern flicker, 90–91

northern mockingbird, 82–83
Nycticorax nycticorax, 150–51

osprey, 158–59
owl, barred, 146–47
owl, great horned, 108–9
oystercatcher, American, 198–99

painted bunting, 16–17
Pandion haliaetus, 158–59
Passerculus sandwichensis, 116–17
Passer domesticus, 32–33
Passerina ciris, 16–17
Passerina cyanea, 14–15
Pelecanus occidentalis, 208–9
pelican, brown, 208–9
Phalacrocorax auritus, 206–7
phoebe, eastern, 40–41
Picoides pubescens, 38–39
pied-billed grebe, 124–25
pigeon, rock, 94–95
pileated woodpecker, 98–99
pine warbler, 26–27
Pipilo erythrophthalmus, 60–61
Piranga rubra, 50–51
plover, black-bellied, 186–87
plover, semipalmated, 178–79
Pluvialis squatarola, 186–87
Podilymbus podiceps, 124–25
Poecile carolinensis, 8–9
Polioptila caerulea, 10–11
Progne subis, 56–57
purple martin, 56–57

Quiscalus major, 196–97
Quiscalus quiscula, 92–93

rail, clapper, 132–33
Rallus longirostris, 132–33

INDEX

217